T0258451

The Alchemy of Disease

The Alchemy of Disease

How Chemicals and Toxins Cause

Cancer and Other Illnesses

John Whysner

Columbia University Press

New York

Columbia University Press
Publishers Since 1893
New York   Chichester, West Sussex
cup.columbia.edu

Library of Congress Cataloging-in-Publication Data
Names: Whysner, John, author.
Title: The alchemy of disease : how chemicals and toxins cause cancer and other illnesses /
    John Whysner, MD, PhD, DABT.
Description: New York : Columbia University Press, [2020] | Includes bibliographical
    references and index.
Identifiers: LCCN 2019043871 (print) | LCCN 2019043872 (ebook) | ISBN 9780231191661
    (hardback) | ISBN 9780231549509 (ebook)
Subjects: LCSH: Toxicology.
Classification: LCC RA1211 .W49 2020 (print) | LCC RA1211 (ebook) | DDC 615.9—dc23
LC record available at https://lccn.loc.gov/2019043871
LC ebook record available at https://lccn.loc.gov/2019043872

Columbia University Press books are printed
on permanent and durable acid-free paper.

Printed in the United States of America

Cover image: © Dorling Kindersley / UIG / Bridgeman Images
Cover design: Chang Jae Lee

This book is dedicated to Paul R. Saunders

Who can be an enemy of alchemy, since it bears no guilt? Guilty is he who does not know it properly and who does not apply it properly.

—Paracelsus

Contents

Acknowledgments

This book is dedicated to Paul R. Saunders, in whose laboratory I was lucky enough to work beginning in high school.
Our studies of the toxicology of venoms, described in chapter 2, was largely responsible for beginning my career. I am also
grateful to Boyd Harding, who was my mentor during my MD-
PhD in biochemistry. My fraternity brother from Johns Hopkins
Ray Milkman convinced Jerome Jaffe to hire me to work in the
Special Action Office for Drug Abuse Prevention, which was part
of Richard Nixon's Executive Office of the President. This connection not only led me to understanding drug abuse treatment
but also provided a connection to my lead-based-paint poisoning prevention studies.

Ken Chase and I were in the same laboratory at the National
Institutes of Health. Ten years later, my employment in his medical and consulting practice Washington Occupational Health
Associates, Inc., provided my transition into occupational and
environmental medicine. I still work for his firm, and the breadth
of my experience in toxicology is largely thanks to his confidence
in my ability to step into any situation and help the workplace
or environment. I am grateful to Gary Williams, who gave me

the opportunity to work in his division at the American Health Foundation when I wanted to return to academic laboratory research. There I met and worked for Ernst Wynder, who helped shape my views of the priorities regarding risks to human health. His historical studies in epidemiology and toxicology showing that cigarette smoking caused lung cancer were inspirational to me. My relationship with the veterinary pathologist Gordon Hard at the foundation was valuable to my understanding species differences for the toxic and carcinogenic effects of chemicals.

After I left the American Health Foundation, Paul Brandt-Rauf offered me a faculty position at the Mailman School of Public Health, Columbia University. Teaching students in his Department of Environmental Health Sciences was a challenging experience: I had never done it before. Teaching the course Fundamental Principles of Toxicology, with the help of fellow faculty members Greg Freyer, Joseph Graziano, and Tom Hei, forced me to round out my knowledge of toxicology.

I am very thankful to my daughter Kate Whysner, who encouraged me to write this book and edited all of it. My other daughter, Joanna Whysner, involved me in global-warming issues through the kid's book that she illustrated, *The Global Warming Express*. My spouse, Amy Bianco, a trade-science book editor, agent, and buyer, provided me with invaluable help in understanding the process of writing a book and getting it published. Also, I wish to thank her brother Anthony Bianco, who is an accomplished business writer, for helping me put together the proposal for this book. I greatly appreciate the colleagues who reviewed book chapters: Jonathan Borak, Christopher Borgert, Paul Brant-Rauf, Frederick Davis, David Eastmond, Joshua Gardner, Joe Graziano, Gordan Hard, Dominick LaCapra, Sonya Lunder, Bob Pearlman, Jerry Rice, Laurence Riff, Ben Stonelake, and especially Sam Cohen, who reviewed multiple chapters.

Special thanks are owed to the editors at Columbia University Press. Miranda Martin and Brian C. Smith have shepherded my manuscript through numerous review processes and provided encouragement for the required reorganization and corrections, which have greatly enhanced the result. Patrick Fitzgerald was my initial contact, who generously entertained my proposal leading to our contractual arrangement. Columbia's production editor Michael Haskell, manuscript editor Robert Fellman, and book designer Chang Jae Lee have greatly enhanced the literary and visual appeal of the book. Finally, I wish to thank the reviewers of the manuscript for Columbia University Press for their valuable criticism and suggestions as well as the Faculty Board and Editorial Committee for their approval.

I acknowledge receiving generous scholarships from the Southern California Alumni for my undergraduate studies at Johns Hopkins University and the American Cancer Society for my MD-PhD studies and research. I have received funding from the National Science Foundation and Department of the Navy for my cone snail venom research; the National Institute on Drug Abuse for clinical studies of opiate addiction treatment; the Department of Housing and Urban Development, the Centers for Disease Control, and the National Bureau of Standards for lead-based-paint poisoning prevention research; and the National Cancer Institute for some of my research at the American Health Foundation. At the American Health Foundation I also received funding from the chemical industry for studies on the mechanism of acrylonitrile-induced tumors in rats, the American Petroleum Institute for studies of the mechanism of benzene-induced acute myelogenous leukemia in humans, Hoffman-LaRoche for DNA binding studies, and the General Electric Corporation for studies on the mechanism of PCB-induced liver tumors in rats.

I disclose that, during my tenure at Washington Occupational Health Associates, I have consulted and provided expert testimony in litigation for the federal government, public utilities, petroleum companies, drug companies, and others regarding polychlorinated biphenyls (PCBs), petroleum solvents, pharmaceuticals, and other chemicals. I have received no funding from the government or industry for the writing and production of this book, and all of the opinions stated herein are mine alone.

Introduction

A s long as humankind has inhabited the earth, nature has threatened us with disease at every turn. Given all of the poisonous plants in our environment, we can imagine the perilous processes of trial and error that early humans must have gone through to find nutritious, safe food and medicinal herbs. From biblical times onward, there are references to eating poisonous foods and drinking poisonous water.[1] Smoke and debris from volcanoes and cooking and forest fires have polluted the air and water. The volcanic eruptions documented in recorded history are only a small part of the episodes of exposures to poisonous gases and particles suffered by ancient humans.[2] Venomous snakes and other animals documented in the Bible would have brought terror, especially to people living in hot climates.[3] And of course there was the constant threat of infectious disease.

As we have gained technological mastery over many of nature's threats, we have come increasingly to fear our own innovations, such as manmade chemicals. We know that we are surrounded by a bewildering array of these threats. We know that they can be toxic and that many cause cancers, but their complexity confounds our attempts to protect ourselves. Nobody wants to get sick from chemicals and toxins. But how can we avoid this fate? How do we identify what can harm us when we usually can't see it, smell it, or touch it? Thousands of scientists are working to give us the answers to these questions. They are called toxicologists, and I have been one of them for more than fifty years.

This book is an account of the history and key findings of toxicology, which began in ancient times and was developed in the Renaissance out of the medieval mysteries of alchemy. Paracelsus, who is considered to be the first toxicologist, was both a physician and an alchemist. Thus, the title of this book, *The Alchemy of Disease*. Alchemy was a complex discipline that included the experimental approach to refinement and transformation of chemicals. However, my account of toxicology primarily involves the industrial age and extends into the information age, during which the major contributions to today's toxicology have occurred. Toxicologists often use the word "xenobiotic" to identify a chemical not naturally produced or expected to be in an organism. In this book the chemicals described are xenobiotics unless otherwise qualified.

Why do we believe that we continue to get sick from chemicals and toxins in our environment? I hope to answer this question in the following pages, and my argument can be summarized as follows: Toxicology has evolved into a science that has made great discoveries over the past five hundred years, and in many cases, the application of this knowledge for public health has been successful. But in other cases, it has been inadequate. The history of toxicology is rich with interesting scientific discoveries and the personalities of physicians, epidemiologists, and laboratory scientists. The abundant discoveries concerning the health effects of chemicals and toxins have provided the information necessary to make decisions about protecting our health historically and currently.

One common thread in this historical description is the difficulty in using animal studies to predict chemical-induced human diseases, especially cancer. Human studies of chemicals are often inadequate or absent, and our current knowledge often relies on animal studies that show toxicity but are of questionable relevance for humans. Another problem that promotes perceived threats from chemicals involves the communication of this knowledge. We have discovered some answers to our concerns, but their meaning is often conditional, which sends a confusing message to the public. We'd like to be able to tell people that chemical X causes disease Y, but such a simple causal chain usually gets compromised by qualifiers: in any given instance it depends upon the disease, the nature of the chemical, and the circumstances and extent of the exposure. These qualifiers are further modified by the current state of our knowledge. Even when we have good reason to think we have established a link between an exposure and an outcome, sometimes it applies and sometimes it doesn't. Addressing legitimate concerns regarding chemical threats is, therefore, not straightforward.

Compounding this dilemma is the need for political and personal willpower to control chemical exposures that have been found to cause harm. This knowledge does not necessarily translate into the appropriate actions, and these deficiencies in motivation are the primary reasons that we may still get sick from chemicals and other agents.

• • •

Toxicology is not the only discipline that contains contradictions and difficulties. I know firsthand that studying medicine can be a humbling experience—I was trained to be a physician as well as a toxicologist. As a medical student I learned anatomy, physiology, biochemistry, pharmacology, and the study of diseases and how to treat them. But I soon found that the difficult part is applying these facts in the real world, on hospital wards or in clinics, to real people who have a disease that we may recognize but often do not. Even if we recognize the disease, we often don't know how to treat it. And sometimes the patient won't cooperate.

One of my earliest professional influences was Roger O. Egeberg, who had been General Douglas MacArthur's aide-de-camp and later became assistant secretary for health and scientific affairs at the Department of Health, Education, and Welfare. Between these two positions, he was dean at the University of Southern California's medical school, where he gave the introductory lecture to my freshman medical school class. He informed us that only since the beginning of the twentieth century had physicians done more good than harm to their patients. Much of medicine before that had been a botched job because physicians were only beginning to understand diseases and their proper treatments. Not only did doctors kill patients with procedures such as phlebotomy, but they also inadvertently poisoned them with drugs.

Likewise, we have been poisoned by our environment. Toxicology has made great strides in certain areas but has failed in others. From a historical perspective, occupational diseases caused by chemical exposures have largely been eliminated, and most environmental exposures to industrial chemicals have been greatly reduced. However, other problems have not been solved. The production and use of fossil fuels still cause ambient air pollution and expose workers to cancer and other diseases. Drug addiction results in overdose deaths, contributes to infectious diseases, and causes intractable social problems. The methods of testing for carcinogens and other health effects using rodent models have never been adequately validated, so we remain uncertain about our ability to detect human hazards. Consequently, this book describes both the successes and failures of toxicology.

I have organized the book into four parts. Introducing the reader to the need for toxicology, part 1 provides examples of environmental problems along with a description of the early history of toxicology, the development of occupational medicine, the evolution of chemical proliferation, and the resulting need for testing chemicals in animals. Chapter 1 explores one major aspect of toxicology: how we determine cause and effect between a chemical and a disease in a specific circumstance. The evaluation of "cancer clusters" in this chapter embraces the challenging question of whether a type of cancer affecting more people than expected in a defined geography can be blamed on a chemical present in the environment. In chapter 2 another aspect is explored: how toxicology shows experimentally how the chemical or toxin causes the disease.

The next two chapters describe the historical development of the basic principles in toxicology. It begins in ancient Greece, but the late Middle Ages, Renaissance, and Enlightenment are emphasized because the knowledge gained during these periods would merge directly with the health problems of the Industrial Revolution.[4] For example, the importance of the dose of the chemical in causing disease was defined by Paracelsus. The identification of occupational diseases by Ramazinni and others laid the historical foundation for toxicology's role in protecting workers. Finally, at the end of part 1 we examine how the explosive development of industrial chemicals and pharmaceuticals caused diseases in workers and consumers, which toxicology strived to understand. The magnitude of the chemical problem and lack of human studies led to the need for toxicity testing in animal bioassays.

In part 2, we discover how toxicologists studied and learned about the toxic and carcinogenic effects of chemicals by studying human diseases and effects in animals. Two of the problems created by the Industrial Revolution are exemplified: poisonings from lead in paint, air, and water and the studies of pesticide exposures described by Rachel Carson. One of the most important aspects of toxicology is the study of how chemicals cause cancer, and in three chapters, I summarize studies of the molecular basis of chemical-induced cancer. Breakthrough studies in humans in the 1950s showed that cigarettes caused the most common form of cancer: lung cancer. Finally, we explore the conclusions of epidemiologists regarding the major causes of cancer.

In part 3, we will see how toxicology is used by society in more practical terms. Toxicological information allows us to regulate chemicals, set the doses of pharmaceuticals that are not toxic, and determine acceptable levels of soil contaminants and airborne chemicals. When these fail, the judicial system uses toxicologists as expert witnesses in lawsuits for claimed injuries in the workplace and environment and for pharmaceuticals and medical devices. Finally, we explore the dark side of toxicology: the development and use of chemical warfare agents.

In part 4, some of the more controversial topics and unfinished business of toxicology are examined, including those areas where toxicology has failed to produce positive public health results. The first of these is opiate addiction and the overdose crisis. The next is the health and climate problems caused by the production and burning of fossil fuels. In three chapters, the problems with animal models for studying and treating human disease are explored. The translation to humans of chemotherapeutic effects on cancer in mice is often disappointing, as is the use of rodent bioassays of chemicals to predict toxicity, including cancer and hormonal effects. This leads to the exploration of alternatives to the conventional rodent bioassays for testing chemicals. Finally, I present suggestions for disease prevention via the application of our present toxicological knowledge to unresolved public health problems.

Although the chapters in this last part describe the areas of toxicology where I believe there is still additional work needed, most of the other chapters in the book also describe topics where we may still be less successful than we would like to be. Childhood lead poisoning serves as an example of an environmental threat that was uncovered more than fifty years ago, thanks to the diligence of many researchers, including toxicologists. After much early success, however, the government dropped the ball, and we have yet to solve the problem completely in the United States. In another chapter we find that many occupational diseases caused by industrial exposures have been greatly mitigated, but new ones are still emerging, such as the lung disease silicosis, which can result from exposure to the large amounts of sand used in hydraulic fracturing for fossil fuel production.

This is not a comprehensive history or textbook of toxicology. Many topics have not been covered, and each topic included here is worthy of a book of its own. My purpose is to provide a narrative of fundamental issues in toxicology and to show the historical development of information in this discipline. Students may find this a good introduction to toxicology before studying the field's comprehensive textbook, Casarett and Doull's *Toxicology: The Basic Science of Poisons*. Although I have attempted to bring these topics to their current state of knowledge, this is not an exhaustive review.

This book is also not a memoir, but I do provide an insider's perspective for many of these subjects. My career has not followed a linear path, so I have participated in a variety of different aspects of toxicology, which I will now briefly describe so that readers don't become confused when I talk about personal experiences in many seemingly unrelated areas. I began studying marine venoms while still in high school and during the summers in college when I attended Johns Hopkins University. This research was done at the University of Southern California School of Medicine with Paul Saunders, to whom this book is dedicated. After receiving my MD and PhD in biochemistry at this school, I studied sea snake venoms at the National

Institutes of Health. Next, I worked for the federal government on opiate addition research and lead-based-paint poisoning prevention. I then became a consultant in occupational and environmental medicine while also working for a research institute studying the cancer-producing mechanisms of chemicals in experimental animals and humans. Finally, I taught graduate-level students in the Department of Environmental Health Sciences at the Mailman School of Public Health at Columbia University. This book is my assessment of the history and current status of toxicology based on my experiences and research. I hope you enjoy the journey.

The Alchemy of Disease

Why Do We Need Toxicology?

Pharmacology and toxicology were one and the same thing—
We were healed by poisons,
And a substance considered an agent of life could,
Under certain circumstances,
In a single convulsion kill within seconds.
—Thomas Mann, *The Magic Mountain* (1924)

The first six chapters of this book introduce the archetypal themes and historical perspectives that justify the need for toxicology. First, we will see how toxicology can contribute to solving the problem of whether chemical exposures cause disease by looking at the phenomenon of "cancer clusters," that is, the unexplained high occurrence of cancer in a geographic area. In contrast to this, in situations where causation is obvious—for example, the sickness and death attributable to a venomous agent—toxicology can explain *how* the toxic effects are caused by components of the venom. In essence, these two contrasting examples serve as the bookends of toxicology: the challenge of determining causation versus the challenge of understanding the mechanisms of toxic effects.

The third and fourth chapters describe the historical development of toxicological methods. Toxicology emerged from the foundations of alchemy and medicine at the end of the Middle Ages by Paracelsus, who is considered the father of the discipline. Later, Ramazinni and others developed occupational medicine by looking at mining and other trades. Next, we will examine the development of synthetic chemistry based on the use of refined organic chemicals and the occupational exposures that produced cancer and other diseases in workers. Part 1 will be rounded out with a discussion of poisonings and of the birth defects and cancers caused by drugs and other synthetic chemicals, the study of which eventually led to the extensive use of experimental animals and eventually large-scale bioassays.

1

Cancer Clusters

Truth Can Be Obscure

Epidemiologists and research toxicologists have learned much about the harmfulness of drugs and other chemicals by studying chemically exposed people and by performing animal experiments or other laboratory tests. But they now appreciate the barriers to translating the knowledge of the toxic effects of chemicals into an understanding of their impact on human health in a given situation. In other words, it is difficult to extrapolate research results to actual people exposed to chemicals in real-life circumstances, and more often than not we are left with a puzzle instead of a solution.

The toxicologist's quest is to find the *toxicon*. The word means "poison arrow" in ancient Greek; it encompasses both the toxin and its vector, or mode of delivery.[1] Finding the toxicon can be relatively straightforward when a person is bitten by a snake; all we have to do is isolate the primary toxic component of the snake venom, and it is evident that the "arrow" is the bite. But most of the time, our target is hiding in a thicket of potentially toxic substances, and it is not clear what form the vector takes.

This difficulty is evident in phenomena called cancer clusters, when more people get cancer than would be expected in a particular workplace or neighborhood. If people in this group are all afflicted by one particular type of cancer—especially if it is a less common cancer—we suspect there may be a single cause. Because some cancers are extremely rare under normal circumstances, sometimes even a few cases of cancer can raise the suspicion that something is wrong in the environment.

• • •

On January 5, 1970, a worker at a B. F. Goodrich Chemical Company plant in Louisville, Kentucky, was hospitalized because he was passing tarry-looking stools, an indication of bleeding in the gastrointestinal system. The worker was initially treated for a stomach ulcer, but on May 1, 1970, he was again admitted with tarry stools. This time palpation revealed that his liver was enlarged, and a liver scan found a large lesion in the organ's left lobe, which can cause bleeding in the veins of the esophagus. An exploratory operation a week later revealed an angiosarcoma of the liver, and the worker died a few months later. The typical type of liver cancer—hepatocellular carcinoma—affects the liver cells, whereas angiosarcoma is a rare cancer of the blood vessels in the liver with an incidence three-hundred-fold less.

The chemical worker was a patient of Dr. J. L. Creech, a plant physician at the company. Two years later, Dr. Creech became aware of two more deaths from angiosarcoma among former workers at the plant who had been treated by other physicians. He connected the dots and learned that all three men who had died of this rare liver tumor had worked at the plant making polyvinyl chloride (PVC) resins from the chemical monomer vinyl chloride. Further investigations revealed that the workers had been exposed to high airborne levels of vinyl chloride in tasks that included entering and cleaning the reactor vessels where the PVC was produced. In January 1974, the manufacturer notified employees at the plant, the National Institute of Occupational Safety and Health, and the Kentucky State Department of Labor of the danger. On February 9, they reported a total of four cases in the *Morbidity and Mortality Weekly Report* of the Centers of Disease Control. When they published their findings in the *Journal of Occupational*

Medicine in March of that year,[2] an editor's note attached to their paper noted that Cesare Maltoni, director of the Institute of Oncology of the University of Bologna, Italy, had just presented, in February 1974, the results of his unpublished experiments finding that vinyl chloride produced angiosarcoma of the liver in experimental animals.

By following up on more medical records, in 1975 Creech, along with Clark Heath Jr. and Henry Falk, from the Centers for Disease Control, reported thirteen cases of angiosarcoma of the liver, including the three cases reported by Creech and Johnson, among men employed at four polyvinyl chloride plants. Data from the National Cancer Institute's Third National Cancer Survey (1969–1971) indicated that only twenty-five to thirty such cases would be expected each year among the entire U.S. population, giving a four-hundred-fold excess.[3] Then, in 1976, another study in a PVC plant found an additional two cases of angiosarcoma of the liver. Subsequent investigations of vinyl chloride workers in the United States, Great Britain, Canada, Sweden, Germany, Italy, and Norway found additional cases of this cancer. These striking increases were enough to indict vinyl chloride as a human carcinogen.[4]

So here we have a cancer cluster identified in a Kentucky manufacturing plant that was quickly confirmed as a major occupational health problem in the PVC industry. Several factors made it relatively easy to find the toxicon in this case: the cancer was a rare type, the exposures occurred in a controlled environment, and the chemical—vinyl chloride—was found to cause the same type of tumor in both humans and experimental animals. All of these combined to provide confirmation of Dr. Creech's initial observations. Other cancer clusters would not be so cooperative.

• • •

Vinyl chloride appears as the accused villain in our next story as well. This time nineteen brain cancers were reported in McHenry County, Illinois. It started with three neighbors diagnosed with brain cancer in 2006, which was an unusual enough event to arouse suspicion over a possible cause. These were not workers in a manufacturing plant but people who lived near chemical plants (owned by Rohm and Haas and Modine) using chlorinated solvents that could degrade into vinyl chloride in the groundwater. It seemed to the residents that they had found an explanation for the high rate of brain cancer in their community. The residents hired an attorney, and eventually he gathered more cases of different types of brain cancers.

Although vinyl chloride had been found clearly to cause liver cancers in humans by organizations such as the International Agency for Research on Cancer (IARC), whose business involves evaluating epidemiology studies, there was no increase in liver cancers among the residents in the area of this cancer cluster. Furthermore, the drinking water of the residents had been tested by the Illinois state government and found to be devoid of any vinyl chloride, and there was no clear pathway of exposure from the sites of the chemical plants to the drinking-water supply.

Gary Ginsberg was a PhD toxicologist with the Connecticut Department of Public Health who had been hired by the law firm representing the plaintiffs in the McHenry County case. He gave the opinion that rats and humans had been found to have brain cancers caused by vinyl chloride and that the cancer risks he calculated for the residents supported their fears that the cancers were caused by vinyl chloride. However, the reported brain cancers in the vinyl chloride studies he cited in the rat studies were actually nasal tumors that had metastasized to the brain. They had a different cell of origin from other types of brain tumors in rats. It was true that the PVC plant in Louisville, Kentucky—the one where Dr. Creech had first found the liver angiosarcoma cases—had also shown an increase in brain cancer, but studies of other PVC plants in the United States and Europe did not confirm this finding. And when the International Agency for Research on Cancer reviewed the studies on vinyl chloride in 1999, it came to the same conclusion that vinyl chloride didn't cause brain tumors.[5]

So what about this cancer cluster? The McHenry County cases were actually five types of brain cancer with different cell types and another benign tumor that is not considered to be a brain cancer. The law firms for the residents also hired an epidemiologist named Dr. Richard Neugebauer from the Columbia School of Public Health, who found that the cancer rate for two types of brain tumors exceeded the expected rate in this community, but the defense epidemiology expert, Dr. Patricia Buffler, who was the former dean of the School of Public Health at UC Berkeley, disputed his findings. She contended that the brain cancer rates were inflated because of the inclusion of nonresidents, the selection of the timeframe, and other factors. But the real kicker was that Neugebauer kept on analyzing the data differently and changed his report even during the trial. Disgusted with the witness, the infuriated judge called the trial off, asserting, "It is as close as I have come sitting on the bench for 20-plus years to having a report that may be tantamount to fraud on the Court, and I will not allow this testimony to continue."[6] The lawsuit was dismissed, but another was filed. Rohm and Hass eventually settled the case in 2014.

• • •

One of the most famous—or infamous—cancer clusters in the United States involved an outbreak of childhood leukemia in Woburn, Massachusetts, in the 1970s. Woburn is located twelve miles northwest of Boston and was a major leather processing and chemical production center in the nineteenth and early twentieth centuries. Beginning in the mid-1960s, residents began to notice a change in the taste and smell of their drinking water. This coincided with the drilling of two new wells that fed the water supply primarily for the eastern part of town, Well G and Well H. In 1967, the Massachusetts Department of Health considered closing the wells because of bacterial contamination, but because the city needed the water, chlorination was initiated instead. This did not improve the aesthetic qualities of the water, however, and angry residents petitioned the city to provide acceptable drinking water. The city took the new wells offline in response, but during drought years it put them back in service, much to the dismay of the consumers. It was not until more than a decade later, in 1979, during the criminal investigation of the dumping of 184 barrels of industrial waste nearby, that the chemical trichloroethylene (TCE) was detected in the water.[7]

Meanwhile, a Boston pediatric hematologist named Dr. John Truman reported that he had seen six cases of leukemia in a six-block area of Woburn beginning in 1972, and a local clergyman, Rev. Bruce Young, reported that ten cases of childhood leukemia had occurred in the eastern area of the town over the previous fifteen years. Residents were convinced that the culprit was the TCE that had been found in their drinking water. They formed a citizens' group and in late 1979 produced a list of local children who had been diagnosed with leukemia. This prompted the Massachusetts Department of Public Health and the U.S. Centers for Disease Control (CDC) to conduct an investigation, and they reported their findings in January 1981.

The CDC conducted interviews revealing that residents of Woburn and neighboring towns had been complaining about the taste of their water and the smell of their air for at least a century. The area had been primarily agricultural until 1853, when land was acquired by the Merrimac Chemical Company to build a plant to produce acids and other chemicals for the textile, leather, and paper industries. In 1899, the company acquired an adjacent plant to produce the pesticide lead arsenate; between 1899 and 1915, Woburn was the leading U.S. producer of arsenical compounds for insect control. The ownership and activities on this and other properties was complicated because parcels were sold and acquired by different entities. Some of the Merrimac property was eventually sold to Stauffer Chemical Company, which produced various chemicals through the 1920s and then animal glues until 1970.[8] Another property was the John J. Riley Tannery, the last of Woburn's twenty or more tanneries, which had been acquired by Beatrice Foods of Chicago.[9] Another property was owned by W. R. Grace, headquartered in New York. They built a plant in 1960 that produced machinery for the food-processing industry, and although the plant did not manufacture chemicals, it was alleged that they used chemicals in their production of machinery.[10]

The CDC investigation confirmed that there was a significantly elevated incidence of childhood leukemia in Woburn from 1969 to 1979. Twelve cases were observed where 5.3 cases would be expected. Most importantly, when the eastern part of Woburn was analyzed separately it revealed a concentration of cases at least seven times greater than expected, while the rest of Woburn exhibited no increase.

Researchers from the Harvard School of Public Health and the Division of Biostatistics and Epidemiology, Dana-Farber Cancer Institute, in Boston confirmed the finding that there was TCE in Wells G and H that served this part of Woburn.[11] At this point it certainly looked like TCE had caused the cancer cases. There was a tremendous outcry in the media, and a famous class-action lawsuit was filed against the corporations W. R. Grace and Beatrice, inspiring the bestselling book *A Civil Action* by Jonathan Harr and a popular movie by the same title, starring John Travolta as the plaintiff's attorney, Jan Schlichtmann, and Robert Duvall as Beatrice's attorney, Jerome Facher. In a controversial decision, the judge divided the trial in two parts, with the first part focusing exclusively on the question of how the water had become contaminated and by whom. The jury held W. R. Grace responsible and let Beatrice go. The second part of the trial was to determine if TCE actually causes leukemia, but because the prolonged and expensive case had financially drained the plaintiffs' attorney, the case was settled before this part of the trial was heard.

After much investigation, the EPA declared the contaminated area a Superfund site and required site remediation of oils and treatment of the groundwater beginning in 1989. They identified five sources of contamination—W. R. Grace & Company, Unifirst Corporation, New England Plastics, Wildwood Conservation Corporation (also referred to as the Beatrice property), and Olympia Nominee Trust.[12]

As for the cancer cluster, the EPA finally concluded that the existence of a true cancer cluster was problematic.[13] And a recent comprehensive review of TCE data by the EPA did not find that TCE causes leukemia, although it does cause kidney cancer.[14] Thus, the initial cancer cluster of childhood leukemia associated with TCE in the drinking water at Woburn did not stand up as a proven causal association, in comparison with the one between vinyl chloride and angiosarcoma of the liver at the PVC plants.

Another childhood leukemia cancer cluster took place in Fallon, Nevada, the current home of the U.S. Navy Top Gun School, also known as the Naval Air Station–Fallon. Top Gun moved from California to Nevada in 1996. In July 2000, an astute local health-care provider notified state health officials that several children in Churchill County had recently been diagnosed with leukemia. This county of only 26,000 permanent residents saw more than a dozen childhood leukemia diagnoses during a four-year period in which fewer than two cases would be expected. This was truly a significant cancer cluster not only because of the cancer incidence increase but also because of the defined and bounded geography: the affected population was not open to interpretation as often occurs in other cancer clusters.

Suspicion focused on a pipeline that runs through downtown Fallon, delivering a continuous supply of JP-8 jet fuel to the base. It was thought that there might be leaks in the pipeline and that the jet fuel might contain benzene, which is known to cause leukemia.[15] This hypothesis quickly fell apart, however, for a number of reasons. No leaks were ever found in the pipeline, and JP-8 is highly refined kerosene with little or no benzene content. Although benzene is known to cause acute myelogenous leukemia, the leukemia cases that were causing the increased incidence were of a type called acute lymphocytic leukemia, which has not been proven to be caused by benzene. Also, researchers at the University of California School of Public Health and the Nevada State Health Division at Carson City, Nevada, looked at other sites where jet fuel was used and did not find any increases in the incidence of childhood leukemia.[16]

Other possible suspects in the cluster at Fallon were naturally occurring high arsenic levels in the drinking water, pesticide exposures, and tungsten in the air from two tungsten-refining facilities. These substances were not known to cause leukemia. Perhaps most tellingly, the Fallon cancer cluster abruptly stopped after four years, though these potential exposures had been present before the cluster began and continued afterward. Researchers from the Centers for Disease Control concluded that even though this was an extraordinary high number of leukemia cases, "Nonetheless, the inability of modern science to identify the role of environmental exposures in leukemia incidence reflects the complexity of defining a relationship between exposure and cancer in a community setting."[17]

It's not known why the kids in Woburn or Fallon got leukemia. Increases of childhood leukemia have been reported to occur after outbreaks of influenza.[18] It is widely accepted that certain viruses can cause certain cancers. Significant excesses of childhood leukemia have repeatedly been found in rural areas that receive a high number of new residents over a short period of time. This has been found in the North Sea petroleum-drilling communities where outsiders came in contact with indigenous Scottish populations. Also, Seascale, on the Irish Sea coast of northern England, experienced such a cancer cluster when the construction of a nuclear power plant brought in workers from outside this rural area. Such rural "mixing" situations tend to promote epidemics because the newcomers are often carrying latent infections to which the original inhabitants have not been exposed. This phenomenon supports the infective hypothesis of childhood leukemia. The initial infection may occur before birth, and the development of childhood leukemia is then triggered later by multiple infectious episodes.[19] The population of Fallon was only about 7,000 when the Top Gun moved there in 1996, and in just a few years it was inundated with about 100,000 military personnel.[20] The origin of the most common form of childhood leukemia is now considered to include infections that occur later rather than earlier in childhood combined with a genetic change that occurs before birth.[21]

It may seem intuitive to look for causes of cancer by investigating apparent increases in cancer among people who share a common residential environment—cancer clusters—but it turns out that this form of investigation has not been successful. In fact, a study of 428 nonoccupational cancer clusters occurring over a twenty-year period found that an increase in cancer incidence was linked with a clear cause in only one case.[22] It is just not a simple matter to figure out whether a chemical causes cancer. And because cancer clusters are statistical events, there is always the possibility that a cluster may be a random occurrence. As improbable as it may seem, if enough cancer clusters are tested, we will come upon a number that occur by chance.

This phenomenon has been called the Texas sharpshooter fallacy. There is a joke about a Texan who fires gunshots at the side of a barn then paints a target centered on the tightest cluster of hits, claiming to be a sharpshooter. This fallacy is an example of the problem of multiplicity testing: when so many statistical tests are run, there are bound to be some positives. In other words, if you do enough statistical tests, some will be positive because of chance.[23]

However, in the case of the Fallon childhood cancers, according to investigators at the UC Berkeley School of Public Health, the probability that this cluster would occur by chance was about 1 in 232 million.[24] Something must have caused this cancer cluster, and it appears that the extreme degree of population mixing is the best hypothesis. So this positive finding is not attributable to the Texas sharpshooter fallacy, but the cause does not appear to be chemicals.

Only in such tightly bound cases of exposure as vinyl chloride in a defined group of workers causing a rare type of liver cancer was it possible to extend a cancer cluster into a bona fide causal relationship. Such occupational studies, that is, systematic studies of workers who have experienced much greater exposures to chemicals than the general public, have provided most of what we know about chemical toxicology in humans.

An early study of cancer clusters by Stephanie Warner and Timothy Aldrich from the Michigan Department of Health and Oak Ridge National Laboratory summarized the contribution and importance of the understanding of cancer clusters succinctly and eloquently: "Cancer cluster investigations have generally been unproductive in terms of etiologic discoveries yet they may have important benefits in terms of public education, allaying public anxiety about environmental concerns and engendering good will toward government agencies."[25] This interesting aspect of cancer cluster studies has not been adequately explored, but this "feel-good" aspect could have both positive and negative consequences for public health.

2 | Death from Arsenic and Venoms

Truth Can Be Obvious

We understand now that even though there is a plethora of industrial chemicals in the environment, it can be difficult to demonstrate whether they cause cancer on a case-by-case basis. However, some chemical poisons and carcinogens are naturally occurring, and some are ubiquitous in our environment. These naturally occurring poisons, for example arsenic, which is present in some drinking-water sources, may also require extensive study to prove that they cause disease. Other naturally occurring toxins are those produced by venomous animals. Intentional poisoning by purified arsenic or venomous stings or bites has an obvious cause-and-effect relationship, compared to the obscurity of the toxin in cancer clusters.

Arsenic is now known to cause several types of cancer, including skin, bladder, and lung. Additionally, there have been some reports of angiosarcoma, the same rare liver cancer found in the vinyl chloride workers described in the first chapter.[1] Arsenic is an element that ranks twentieth in abundance in the earth's crust, fourteenth in seawater, and twelfth in the human body. The average soil level of arsenic in the United States is 7.5 parts per million. Arsenic can exist in different inorganic and organic forms, the latter being much less toxic. Low levels of mostly organic arsenic are commonly found in food; the highest levels are found in seafood, meats, and grains, especially rice. Of greatest concern for toxicology is inorganic arsenic in wellwater, which is caused by its occurrence in bedrock and volcanic-rock deposits.[2] The chemical form of arsenic in groundwater in Bangladesh and elsewhere in Southeast Asia is arsenite, produced by the reducing conditions in groundwater. (Basically, aerobic bacteria in the sediments consume all of the dissolved oxygen, thereby creating a reducing environment.) In Chile, the arsenic present in drinking water is arsenate, an oxidized form, because drinking water is taken from river water (with mining-related contamination), which is exposed to air. Arsenite's toxicity is greater than arsenate's, which has important implications for the discovery of arsenic in wellwater in Bangladesh.[3]

Usually we think about inorganic arsenic as an acute poison that has a long history of being used for wicked purposes. The Greeks described arsenic poisonings, and during the Renaissance the Borgias in Italy were infamous for killing political opponents with arsenic. Forensic toxicology had its beginnings with the analysis of arsenic developed by James Marsh in 1836. His test was used by the celebrated forensic toxicologist Mathieu Orfila in a famous case to prove that the death of Charles LaFarge was from poisoning by his wife, Marie.[4]

However, for our purposes, the most important aspect of arsenic toxicity involves chronic long-term exposures to small amounts. The first medical finding identified in communities with high arsenic levels in drinking water was a noncancerous dermatological condition discovered at the beginning of the 1960s especially in children of Antofagasta, Chile, a city of 130,000 inhabitants.[5] Separately, in 1961, Wen-Peng Tseng described blackfoot disease, an endemic disorder confined to a limited area on the southwestern coast of Taiwan. Blackfoot disease was a folk term for a peripheral vascular disorder resulting in gangrene of the extremities, especially the feet. The disease, which commonly ends with spontaneous amputation of the affected extremities, was attributable to the water drawn from artesian wells in the area, which were dug by public health officials to prevent the cholera caused by drinking the region's surface waters.

Tseng later observed a high incidence of skin cancer in the same communities where blackfoot disease had been found.[6] This was in essence a newly observed cancer cluster; skin cancer was relatively uncommon in Taiwan. Subsequent investigations by the Institute of Public Health, National Taiwan University College of Medicine, found that not only was skin cancer associated with the dose of arsenic consumed in the drinking water, but so were lung and liver cancers.[7]

The most extensive contamination of inorganic arsenic in groundwater was found in the 1990s in Bangladesh.[8] Cholera had been rampant in Bangladesh because the surface waters used for drinking were contaminated with human waste. In the 1970s the World Bank and UNICEF provided funding for the installation of 10 million tube wells to tap the groundwater, which was clear of pathogens. At that time, arsenic was not recognized as a problem in water supplies, so no tests were performed for this poison. However, arsenic had collected in this groundwater from natural arsenic-rich material, delivered by the region's river systems over millions of years. Once high levels of arsenic were discovered in the wellwater, the people of Bangladesh were caught between two terrible alternatives: a quick death from cholera in surface water or a slow death from arsenic in groundwater.

The contamination of groundwater by arsenic in Bangladesh became the largest poisoning of a population in history, with tens of millions of people exposed. In 1983, the first cases of arsenic-induced skin lesions were identified by K. C. Saha, then at the Department of Dermatology, School of Tropical Medicine in Calcutta, India. The first patients seen were from West Bengal in India; the characteristic skin lesions were keratosis of the palms and soles of the feet, with discoloration of the skin. By 1987 several cases were identified who came from Bangladesh, which borders West Bengal. Arsenic contamination of water in tube wells was confirmed in 1993 in the Nawabganj district, and results from various laboratories were collated in a WHO country situation report in 1996 showing extensive arsenic contamination.[9]

Chile, Taiwan, and Bangladesh are not the only places where high levels of arsenic have been found in the drinking water; other substantial areas with high arsenic levels are parts of northern China, Cambodia, Vietnam,

Argentina, Mexico, and the United States. For example, in the United States, about 3 percent of wells used for drinking water in Arizona have been found to have levels of arsenic that would increase the risk of cancer, according to Samuel Cohen, a widely published toxicologist from the University of Nebraska Medical Center. He determined that drinking-water levels above 100 micrograms per liter can be associated with increased cancer risks.[10] Household wells have also been found to contain high levels of arsenic in parts of California, Texas, New Hampshire, and Maine.

• • •

So far in these first two chapters, we have examined the occurrences of diseases where extensive studies were required to establish a causal link to an environmental contaminant. However, this is not always the case in toxicology. In order to bookend the scope of this discipline, we will now examine the other extreme: agents whereby the complete opposite is true. In the case of venoms produced by snakes and other land or sea creatures, the cause-and-effect relationship of poisoning is obvious. Exposure through a bite or sting from these creatures can cause sudden death. As opposed to the study of cancer clusters, toxicology's methods are much more direct and focused on finding out how these agents cause their effects. The toxicologist is faced with an entirely different question: how do these venoms kill?

The longest short walk of my life came early one morning in about 1964, when I went to meet a physician visiting from Australia to learn a new separation technique for the proteins in venoms. He had to go on 8:00 a.m. rounds at the hospital, so I was to be there at 7:00 sharp to learn about the new polyacrylamide gels. He was visiting Dr. Findlay Russell's laboratory at the University of Southern California. I opened the door to the building where Russell's laboratory was located and found that it was pitch dark inside except for a line of light around a door in the distance about a hundred feet off. As I crept down the long corridor toward the lit door, I heard Sssss-Dt!, Ssss-Dt!, Ssss-Dt!, Ssss-Dt!, Ssss-Dt! all around me. I was a bit panicked by these weird sounds but had no idea what they were. Nervously, I opened the door, and there was my host, who welcomed me warmly. After exchanging greetings, I asked him what the strange noises in the hallway were. He laughed and turned on the lights. To my amazement and a bit of horror, the hall was lined with wire cages stacked five high containing rattlesnakes.

Findlay Russell was a physician who served as a professor of neurology, physiology, and biology at USC for more than thirty years. He was the world's leading expert on rattlesnake venom, and he wrote the chapter on venoms and toxins in *Casarett and Doull's Toxicology*, the field's major textbook. He was also on call to treat people for rattlesnake bites with antivenoms and supportive measures. Los Angeles County Hospital was right across the street from his lab, and it had one of the most active emergency rooms in the United States. Toxins are produced by many living things, and venoms are toxins made primarily of proteins that are injected into the skin either by a bite or sting. (A side note: "poison" is a more general term than "toxin"; the former term includes manmade chemicals, naturally occurring minerals, and other substances that cause harm. However, things are often referred to as being "poisonous" or "toxic" interchangeably.)

Studying the effects of venoms on humans is much more straightforward than the kind of toxicology needed to study a cancer cluster. In general, venoms work quickly, and their effects are obvious. No statistics and no comparison groups are necessary to figure out cause and effect: someone gets bitten or stung, they get sick almost immediately, and sometimes they die. The use of standardized cancer rates is not required to give the epidemiological clues to begin the toxicological research on the effects of venoms. Studies of the pathways of exposure are not necessary. The stings or bites are usually obvious. These biologically active molecules causing the rapid onset of toxicity are at the other end of the discipline's spectrum from exposures to carcinogens whose deleterious effects can take years to manifest.

Unlike cancer, venomous bites are not a major public health concern. But they are an obvious cause of disease and are preventable or at least treatable. Toxicologists have contributed significantly to understanding how venoms cause sickness and death. There is a public awareness, developed over centuries, of the types of animals that can kill. Today, bites from snakes or stings from various animals are no longer the death sentence they used to be, although stings from jellyfish in the South Pacific have recently raised concerns.[11] Our ability to provide supportive care through artificial respiration and the maintenance of blood pressure and cardiac rhythm has greatly decreased the lethality of venomous animals. And we now have antivenoms, which are produced by injecting venom into animals in increasing doses and then extracting the antibodies to the venom from their blood. Emergency rooms everywhere now stock commercially available antivenoms, so they no longer need someone like Findlay Russell to run across the street and save the patient.

• • •

The toxicity of marine life was not always as well understood as it is today. In the 1950s, toxicologists were trying to figure out the basics of how venoms worked. Venoms are complex mixtures of biologically active molecules, and they combine various properties to increase their effectiveness. There are components of the venom that enhance its spread in the tissues by breaking down proteins and preventing blood from coagulating. The main components of most venoms are neurotoxins that either cause paralysis or impair the pumping action of the heart. Much of this work, as well as the development of antivenoms and treatments, was occurring in Southern California. In 1960, I began working for Dr. Paul Saunders, who was the associate dean for medical education at USC and a good friend of Russell's. Saunders studied venomous marine organisms such as the stonefish and cone snail. Together, he and Russell founded the International Society of Toxicology and published the journal *Toxicon*. The Office of Naval Research of the Department of the Navy funded his research because cone snails were a threat to people on the Pacific Ocean beaches where naval personnel were stationed. Cone snails don't jump out and bite people, but people would pick up these shells, which are very pretty and colorful—not knowing that they were alive—hold them in their hands, and get stung. The cone snail's venom is powerful and can be deadly.

Cone snails are fascinating marine creatures that kill their prey by harpooning them. A cone snail will crawl along the ocean floor, sneaking up on its victim and sticking it with a hollow barb to inject its venom. This barb is like a hypodermic needle and is called a radular tooth. It sticks out from the end of the proboscis, a tongue-like structure that protrudes out of the snail's mouth. A muscular bulb inside the snail provides the force to inject the venom. In between the muscular bulb and the radular tooth is the venom duct, where the venom is produced. Once it has rendered its victim helpless, the cone snail draws its meal into its mouth with its barbed proboscis.

Saunders had living cone snail specimens in tanks, which he kept alive as a curiosity. They had come from the South Pacific near the Great Barrier Reef and were finicky about what they would prey upon. Goldfish were not good enough, so to keep these cone snails alive and happy I collected live food from tide pools at the ocean shore. The *Conus geographus*—the largest species of cone snail—and *Conus straitus* would eat only fish, but the *Conus textile* dined on other snails. Many visitors, including scientists from around the world, would come to watch them eat and film movies of them to show their students. Black turban snails, *Tegula funebralis*, were abundant in the tide pools, and the cone snails ate them greedily. They would sneak up on their prey, stick their harpoon-like proboscis into a turban snail's foot, the venom would kill it, and the cone would pull in its feast.

Finding the right prey for the fish-eating cones was trickier. If the fish were dead, the cone snails were not that interested and sometimes would not eat. And the visiting scientists and photographers wanted to observe live action, not mere feeding. A relative of the scorpion fish hovers at the bottom of the tide pools, and once in the cone snail's tank it obliged by lying quietly on the bottom so the snail could creep up on it, harpoon, paralyze, and eat it.

The South Pacific cones were too valuable for Saunders to study in the lab because they could not be milked as snake venoms can. Therefore, the only way to study the venom was to hammer the shell open and dissect the venom duct, which would kill the snail. Fortunately, there is a native variety in Southern California, *Conus californicus*. Oddly, there were no reports of people being stung by this cone snail, despite my eventual finding that this species has a powerful venom. These creatures were the drab cousins—small and black—of the cone snails found in the South Pacific. It makes sense, then, that people would not get stung by them: they are not something one would want to collect. Also, they populate the mucky, grassy bottoms of ocean bays where people don't normally go beachcombing.

Fortunately, I found lots of them in Morro Bay, California, so I was able to do my studies.

When faced with venom from a new species, such as from the cone from Southern California, toxicologists first measure its lethality by determining how much of the venom it takes to kill an animal. The most basic dose-response test in toxicology is called the LD50, which identifies the dose of a venom (or any chemical) required to kill 50 percent of a set of animals. The venom can be injected into the tail veins of groups of mice in various doses; one increases or decreases the dose given to the subsequent group in a stepwise fashion depending on the result in the previous animals. It turns out that cone snail venoms are very powerful, and even the diminutive Californian relatives of the South Pacific cones contain almost as much killing power as the largest fish-eating types or the common rattlesnakes found in the United States.[12]

Even more deadly than the deadliest of cone snails was the stonefish, or *Synanceja horrida*. A stonefish looks like a stone partially buried in sand, but it has spines on its back containing sheaths of venom. If a barefoot person steps on a stonefish, they usually die soon afterward. Stonefish venom is about ten times more powerful than that of the cone snail.[13]

• • •

In the 1950s, toxicology was studied in pharmacology departments. But whereas pharmacology studied the good effects of drugs (efficacy), toxicology studied the bad effects. In those days, most studies in toxicology involved the toxic effects of pharmaceuticals, but Saunders was using the techniques to study venoms. We set out to study how my *Conus californicus* venom worked on anesthetized rabbits, monitoring for blood pressure, heart rate, and respiration. The cone venom caused an immediate decrease in blood pressure and heart rate. Respiratory rate increased, and there were no changes in heart rhythms. At much higher doses, death was preceded by changes in the heart rhythm.[14] These studies were published in 1963 in the third issue of Saunders and Russell's *Toxicon*.

The active components in venoms are proteins, and after determining lethality, the next step is to separate out the various proteins in the venom and find the lethal components. A problem with studying proteins in a laboratory is that exposure to high temperatures causes their inactivation. This meant we had to conduct column chromatography and electrophoresis in the cold room, which is like working in a meat locker all day. The polyacrylamide gels that later became a fundamental tool of molecular toxicology were run at room temperature. This inactivated the effects but was better at showing all of the venom's components. These methods use the electrical charges on the proteins to separate them either by binding them on an inert matrix or by applying an electrical field to the gel, across which the proteins then migrate at different rates. In column chromatography, various solutions are applied that change the ionic properties of the column so that the protein components of the venom first attach to the column matrix and then are released into the liquid flowing through the column at various times. After collecting the proper fractions eluting from the column, the solution had to be concentrated. These techniques showed that there were more than twenty protein components in cone snail venom.[15]

The lethal components of most venoms are neurotoxins. They work by binding to neurotransmitters and rendering them inactive. For example, the chemical acetylcholine is a neurotransmitter through which some nerve cells communicate signals to other nerve cells. Acetylcholine is released from the axon of one nerve cell following an electrical signal, the chemical travels across the synapse—the gap between nerve cells—to bind to a receptor on the next nerve cell, and this causes the next nerve cell to fire off another electrical impulse. Acetylcholine is also the chemical that communicates between nerves and muscles, causing muscles to contract. Scientists from France and Taiwan studied the venom of *Bungarus multicinctus*, also known as the many-banded krait. This deadly marine snake is found in marshy areas in Taiwan and southern China. The lethal component in *Bungarus multicinctus* venom is called α-bungarotoxin, and it binds irreversibly to the acetylcholine receptors, which causes permanent paralysis in all the muscles of the body. In live animals, including humans, the result is sudden death by asphyxiation.[16]

These venoms are a useful tool in medical research. Douglas Fambrough of the Department of Embryology, Carnegie Institution of Washington in Baltimore, and H. Criss Hartzell, of the Department of Biology at Johns

Hopkins University, used the purified α-bungarotoxin bound to radioactive iodine to identify and count acetylcholine receptors in the muscle of the rat diaphragm.[17] This radioactive bungarotoxin could be located using a photographic technique called autoradiography. The α-bungarotoxin binds so irreversibly and specifically to the acetylcholine receptors that one can visualize the synapses in tissue cultures of the spinal cord and the brain. The number of acetylcholine receptors can be counted by measuring the amount of radioactivity attached to cells using another technique called liquid scintillation, which can quantify the number of venom molecules attached to the receptor.

Thus, in the case of venoms, toxicologists do not need to establish a causal relationship with disease, because it is usually obvious. The studies of toxicologists are aimed at understanding the mechanisms of toxicity in the laboratory using animals, cells, and biochemical techniques. This is in stark contrast to studies of cancer clusters caused either by industrial chemicals or by a naturally occurring chemical like arsenic, where a relationship to exposures is often not clear. In addition, the study of venoms has provided clues to the mechanism of certain neurological conditions and even some cancers such as neuroblastoma.

3 | Paracelsus

The Alchemist at Work

To understand the history of toxicology, we should begin by seeing how alchemy influenced the practice of medicine during the Middle Ages. Alchemy is primarily concerned with chemical transformations, in contrast to ancient metallurgy, which dealt with the extraction of metals from their naturally occurring ores. Perhaps the most well-known activity of alchemists is their attempts to "purify" and transform base metals such as lead into precious metals such as gold. Alchemists also sought potions that would give eternal life or could act as universal solvents. In the European tradition of alchemy, the practical chemical work in the laboratory was blended with a psychological and mystical content involving Christian teachings. Whatever the aims of the alchemists were, their trade was steeped in esoteric thought. Today, the hundreds of procedures and recipes developed in the Middle Ages by alchemists have little recognizable meaning for a modern chemist.[1] Nevertheless, some of the procedures and findings led to basic concepts in both chemistry and toxicology.

Philippus Aureolus Theophrastus Bombastus von Hohenheim, also known as Paracelsus, was born in Switzerland in

1493, one year after Columbus landed in Hispaniola and forty years after Gutenberg printed his first bibles. Both these events would be important to Paracelsus's success. The explorers brought back syphilis, which he would treat, and the printing press would allow the wide dissemination of his ideas. Paracelsus is considered the father of toxicology both because he identified environmental causes in the etiology of disease and because he was the first to understand that the dose of a chemical is a primary factor in its adverse effects. Other physicians, going as far back as Hippocrates (460–370 BC), had understood some aspects of the toxicity of drugs and other chemicals, but before Paracelsus, physicians considered remedies to be something apart from poisons. Paracelsus was the first to understand that they could be the same—that therapeutic drugs could also be poisons, that "the dose makes the poison."

Paracelsus's father christened him Theophrastus after the Greek philosopher Tyrtamus Theophrastus of Eresus of Lesbos (371–287 BCE), a successor of Aristotle (384–322 BCE). The name was doubly apt because Theophrastus is considered the father of botany, and we can assume that Paracelsus studied his works to identify medicinal herbs. Theophrastus inherited the school of Aristotle called the Peripatetic School, and this certainly fits another of the notable traits of Paracelsus: he was a wanderer.[2]

Paracelsus obtained a baccalaureate in Vienna, Austria, and like other learned men of his time, he studied alchemy, astrology, religion, and the classical sciences. He then studied medicine in Ferrara and at a number of other universities in Italy. Around that time he assumed the name Paracelsus, meaning "surpassing Celsus," an early Roman physician who wrote the encyclopedic text for physicians *De Medicina*.[3]

Medicine in Paracelsus's time was based entirely on the work of earlier physicians: Hippocrates, Galen, Avicenna, and Razes. Galen, a Greek physician who lived in Roman times, had modified the ancient Greek physician Hippocrates's teachings based upon his study of animals. These teachings were later refined by Muslim physicians.

Hippocrates interviewed patients to systematically connect the assumptions about their illness. He asked about the symptoms of their illness, about the patient's behaviors and conditions before the onset of the illness, and about their residences, diet, and physical activity. In other words, he developed the art of taking a medical history and used this information to develop disease causation.[4] He also made certain ecological comparisons based on geographic differences, seasons of the year, and ethnic groups in order to determine what might be causing diseases among populations. By today's standards, the theoretical aspects of Hippocrates's medical teachings were almost entirely wrong, but he was systematic. He developed a theory of four humors, blood, phlegm, black bile, and yellow bile, whose lack or overabundance was hypothesized to be responsible for health or disease.[5]

Galen used Hippocrates's theory of the balance of the four humors as the basis for his teachings. His understanding of the human body was inferred from animal studies, and when he couldn't ascertain certain facts directly, he filled in the gaps with theoretical constructs based on Platonic ideals. Based on his studies, he formulated remedies in the form of plant extracts. His construct of anatomy and physiology, developed during the second century CE, was so comprehensive that it would dominate the next thousand years of medicine. As a consequence, during the time of Paracelsus, the use of remedies was still rooted in the humoral theory of Hippocrates and Galen, and the medical texts of the time were arcane and complex, discussing an endless variety of pills and potions that could be used to restore the humoral equilibrium.[6]

• • •

Despite the total reliance of the leading academic physicians of his time on Galen and other influences, Paracelsus chose to gain his knowledge of disease firsthand by treating patients. He decided that one must test for himself any knowledge acquired from other physicians, alchemists, shamans, herbalists, and healers. He was in constant motion, diagnosing, treating, studying, and teaching in Germany, Italy, and France.[7]

In 1515 he embarked on a trip into the midst of armed conflicts, acting at times as a military surgeon. The king of Denmark and Norway, Christian II, had named him royal physician. When a group of the Teutonic Knights besieged Danzig and were defeated, Paracelsus traveled with them to Konigsberg, in Prussia. The grand master of the order of Teutonic Knights in Prussia wanted to form a truce with Grand Duke Vassily III of Moscow, the ruler of all Russians, so Paracelsus traveled as an unofficial ambassador to Moscow, which was under siege by the Tartars. Paracelsus was captured by the Tartars, who considered healers holy men and so spared him. Paracelsus continued to teach and practice medicine even as a prisoner in the midst of people he likely considered barbarians. He spent years studying their herbal remedies, and finally in 1521, he was liberated from the Tartars by a troop of Polish knights.[8]

After some time spent back in Venice, he departed for a trip up the Nile River, following a Venetian trading route and other pilgrimage routes. He crossed the Red Sea to the Gulf of Aqaba and spent time in Jerusalem and Ottoman-controlled Greece. In 1523 he wound up in Constantinople, where he was rumored to have learned the secret of alchemical gold making from a German alchemist named Solomon Trismosin. At that time the Ottoman court was considered a center of knowledge, and although Islam was at war with Christendom, it was a dignified war that observed courtly rules, unlike our modern-day conflicts. From Constantinople, Paracelsus embarked on the dangerous journey back to Venice in 1523 and over the Alps to his father's home in Villach, Austria.[9]

The point in recounting the travels of Paracelsus is to demonstrate his thirst for firsthand experience, which made his contribution to toxicology so important. Paracelsus insisted on teaching the skills and knowledge that he personally had gained on his travels rather than relying on the dogma of his predecessors Hippocrates and Galen. This made him a reformer. "I am *monarcha medicorum*, monarch of physicians," he proclaimed, "and I can prove to you what you cannot prove."[10] Paracelsus scorned physicians who were "cushion sitters" and "seat themselves in the midst of books, and ride thus the Ship of Fools." He believed that it was only through constant traveling that knowledge could be gained because specific attributes of each region shaped that region's medicine.[11]

Paracelsus mockingly described the stereotypical medical doctor of the time: "He begins gradually and slowly to doctor him, spends much time on Syrups, on Laxatives, on purgatives and oatmeal mushes, on barley, on pumpkins, on Melons, on Julep and other such rubbish, is slow and frequently administers enemas, does not know himself what he is doing, and thus drags along with time and gentle words till he comes to the term."

Paracelsus also judged doctors for their corruption: "But since in medicine such useless folk are involved who seek and consider only profit, how can it be that I admonish them to love, or how can it have any effect: I for my part am ashamed of medicine, seeing to what utter deceit it has come."[12]

• • •

Alchemy in the Middle Ages had become intertwined with the Catholic Church, which offered analogies connecting the alchemical process to the historical figure of Christ the Redeemer, which therefore detached alchemy from the world of natural phenomena. Paracelsus represented the alchemist's emergence from the medieval concepts of the Catholic Church; he ran counter to its reliance on faith and preferred to seek knowledge through a direct experience of natural processes. Nevertheless, the alchemists of this period still found harmony between the scientific and the divine; Paracelsus and his contemporaries never thought of themselves as anything other than good Christians.[13]

Although he never left the Catholic Church, Paracelsus followed his own beliefs, insisting that one should shun all doctrines except that of the simple truth in the Bible. His religious beliefs are more accurately described as a type of unorthodox, mystical pantheism, however: he held that in addition to his visible body man has an invisible body that interacts with the occult forces of nature.[14] Because of his unorthodox views about medicine, Paracelsus was often compared to his contemporary Martin Luther (1483–1546 CE). But Paracelsus rejected this analogy. In *Das Buch Paragranum*, which he wrote in 1529–1530, Paracelsus protested: "With what scorn you have proclaimed that I am the Luther of physicians, with the interpretation that I am a heretic."[15]

Despite his distaste for the comparison, he took a cue from Luther, who burned the papal bull issued against his teachings. Paracelsus publicly burned the bible of learned medicine, Avicenna's *Canon*, along with other texts based on Galen's teachings.[16] Each St. John's Day, the city of Basel would build a bonfire for effigies of unpopular figures to be burned, and Paracelsus took the opportunity to destroy these works on June 24, 1527, surrounded by his students, so that, as he put it, "all that misery might go up in the air with the smoke."[17] Paracelsus opposed the ancient humoral medicine; he did not believe that imbalances among the four humors could account for the diversity of diseases he had encountered. Rather than trying to cure illness by readjusting the body's humors using treatments such as bloodletting, which was supposed to balance the blood humor, Paracelsus used chemical remedies. His vast experience had convinced him that disease was directly caused by external factors, not internal imbalances; therefore, external cures and remedies could affect the body directly and produce specific results without a need for humoral readjustment.

Paracelsus's alchemically based approach to treatments was in marked contrast to the many hundreds of procedures and recipes used in the Middle Ages and antiquity. Paracelsus found most of the herbal remedies of the Galenists nonsensical. There was seldom any indication of what was really done or used in these experiments, and the names of ingredients could mean almost anything.[18]

In his first great work, *Archidoxa*, or "Arch-Wisdom," he was not concerned with making gold; he was making medicines. Paracelsus's work represents the beginning of chemotherapy, whereby a disease is treated via the direct effect of a chemical on the body rather than by adjusting the internal balances of the body. Though the work of many alchemists was to "kill" materials in order to gain perfection, as in the production of gold, for Paracelsus the underlying fundamental principle of alchemy was that the processes in the microcosm of the living body were reflected in the macrocosm of the universe. For Paracelsus, this meant that the processes in the outer world, including those in the laboratories of alchemists, provided insight into the workings of the human body.[19] Philosophically, alchemy was a process of chemically separating the waste of mundane reality from the vital, healing forces of nature, thereby yielding the rarefied, pure essences of nature.[20]

• • •

Syphilis broke out in 1494 among French soldiers garrisoned in Naples. It spread quickly to Bologna in 1495, and in the following year Swiss soldiers took it home to Geneva. From there it traveled quickly throughout France and Germany. This devastating venereal disease had never been seen before in Europe. Paracelsus believed that syphilis was a transformed version of an older indigenous disease needing a new remedy, but it was probably a new disease for Europe and needed a new remedy. The Germans called it the French disease, but it is believed that Columbus's sailors probably brought it home with them from the West Indies, according to a Spanish physician reporting from a Spanish port.[21]

Although primary syphilis forms a chancre at the site of infection and can be seen on the penis, secondary syphilis is often an extensive rash evident weeks after the primary infection. Thus, since the visible manifestations of syphilis were in the skin, and since Arab physicians several centuries earlier had used mercury for some skin diseases, this toxic metal became a form of treatment for syphilis. The side effects of mercury were horrific: patients suffered uncontrollable tremors, sores inside the mouth and on the tongue, loss of teeth, and frequent, stinking urination.

Because of the toxicity of mercury, an alternative herbal remedy gained popularity for the treatment of syphilis. Sailors found a dense wood in the West Indies called guaiac, which we know as lignum vitae, and an antisyphilitic cure was concocted in the belief that a local disease has a local cure. Shavings of this wood were boiled, and the resulting frothy scum was applied to the sores on the skin and imbibed by the patient. A profitable market developed for this rare commodity. The enterprising Fugger family of Augsburg, Germany, who were prominent European bankers, received a monopoly from Charles I of Spain in return for a loan, which he used to bribe his way to becoming Charles V of the Holy Roman Empire. Unfortunately, the cure did not work on Charles.[22]

Paracelsus made a study of the guaiac treatment of syphilis and became convinced that the cure was worthless. In doing so he was challenging not only the medical establishment but also a thriving commercial enterprise of powerful and influential people. He also understood that the essence of his medicines was their poisonous nature. According to his alchemical philosophy, by purifying these poisons, he would be able to transmute them into remedies, which when administered in the correct formulation, dosage, and regimen would be effective treatments for disease.

Following his alchemical instincts as a student of metals, he focused on the problematic mercury treatment. Moreover, believing that "like cures like," he reasoned that a metal that causes illness by poisoning could treat an illness at the correct dose. In fact, many of the symptoms of mercury poisoning were similar to those of tertiary syphilis. Mercury was clearly toxic, but Paracelsus became convinced that one did not need to administer it at high doses to cure the disease. He found that above a determinable threshold, increasing the dose did not produce any further therapeutic benefit but did increase the toxic effects. Paracelsus found that at the right dose mercury was effective in treating syphilis but did not cause severe toxic effects. This was a great breakthrough in medicine based on the understanding of dose and toxic response. As a consequence, his mercury formulations based on his determination of dose remained the only available treatment of syphilis until the twentieth century, when another metal-based remedy, one involving arsenic, was discovered.[23]

Because of the theory of like treating like, Paracelsus has been at times considered the originator of homeopathy. The concept of dose is also important in homeopathy; however, this comparison is superficial because the doses used in homeopathic remedies are so small that they are close to no dose at all. They neither would have had the therapeutic effects nor induced the near-toxic effects of his remedies.[24]

Paracelsus also used other mineral substances, such as lead, antimony, sulfur, copper, and arsenic, in his treatments.[25] Paracelsus developed a protocol for transforming the deadliest poison into a remedy by altering the dose and carefully administering it with correct timing and for the appropriate disease. As he used mercury to treat syphilis with success, he used other metals as well, such as antimony.[26] He prepared "butter of antimony" as a palatable form of this metal so it could be used as a remedy. He taught that one aim of medicine should be the preparation of what he called "arcana," which are mostly minerals, as opposed to herbal remedies derived from Greco-Roman teaching. Paracelsus used sulfuric acid to treat epilepsy, syphilis, dropsy, gout, and miners' disease.[27] Far from being poorly defined imbalances of purported internal humors, he claimed that diseases were specific entities and could be treated by precisely selected arcana, which would destroy the poisons produced by disease. The role of the alchemist-physician was to separate the pure essence from the impure using the tools of the alchemist: fire and distillation.[28]

The ultimate success of Paracelsus's treatments depended upon his understanding of the toxicology of the remedies that he proposed. He believed that each poison contained a *toxicon*, a primary chemical entity responsible for its toxicity.[29] This concept relied heavily on alchemy, which he conceived of as involving a separation process that removed the impure from the pure. Techniques such as distillation were used by alchemy to concentrate the potency of toxic materials.[30]

The concept of dose also relied on the principle of alchemy: "What the eyes perceive in herbs or stones or trees is not yet a remedy; the eyes see only the dross," Paracelsus wrote. "But inside, under the dross, there the remedy lies hidden. First it must be cleansed from the dross, then it is there. This is alchemy, and this is the office of Vulcan, he is the apothecary and chemist of medicine."[31]

• • •

The legacy of Paracelsus includes the concept that experimentation is essential in the study of the toxicological and pharmacological responses to chemicals. Specificity of response in these experiments is the key to understanding the difference between a therapeutic and a toxic effect. This guided his understanding of the distinction between the therapeutic and toxic properties of chemicals, and these properties are usually distinguishable by dose.[32]

Although he was systematically scientific, Paracelsus was also very religious. He believed that God had given poisons to man and that poisons should not be rejected, because they may also have curative values. He urged physicians not to be frightened by remedies that could also be poisons. He famously stated that "All things are poison, and nothing is without poison: the dose alone makes a thing not a poison. For example, every food and every drink, if taken beyond its dose, is poison: the result proves it. I admit also that poison is poison: that it should however, therefore be rejected, is impossible."[33] Of course, this is a familiar concept to us now. We know that drugs need to be taken in the prescribed amounts and that if we take too much we can suffer the consequences. Paracelsus sought to prevent the hazards of medicines by carefully prescribing the correct dose. Therefore, not only is Paracelsus considered to be the father of toxicology, but he was also the most prominent early proponent of chemotherapy and the originator of the modern medical discipline of pharmacology. [34]

Once considered a heretic within his profession for rejecting a rigid adherence to classical medical thought, Paracelsus would in the end have a great influence on the course of medical science and the treatment of diseases.[35] Paracelsan physicians became respected, and Paracelsanism had great appeal partly because of its effective use of chemotherapy and partly because the devout Christian nature of Paracelsus's teaching appealed to the religious beliefs of the time. Even Puritans praised Paracelsanism for unmasking the heathen roots of Galenism.[36]

4

Mining and the Beginnings of Occupational Medicine

Some occupations are hazardous, and some can even be deadly. Logging workers and commercial fishermen have the most statistically dangerous jobs today, but miners probably have the longest and most infamous history of occupational deaths (mortality) and disease (morbidity).[1] Mining is an industry that has been around since the dawn of civilization, with workers that have always been, and still are, vulnerable to accidents and chemical exposures. Much of our knowledge in toxicology comes from the study of occupations like mining because workers are exposed to chemicals and other agents at high doses and for long periods of time.

The dangers of mining have been recorded since ancient times. Lead mining in Cartagena, Spain, exploited by Carthage and later by the Romans, employed forty thousand men and left a legacy of early disease and death. The ailments specific to lead miners were described first by the Egyptians, who used slaves in their mines, and later by the Greeks and Romans. Hippocrates described the symptoms of lead poisoning as appetite loss, colic, pallor, weight loss, fatigue, irritability, and nervous spasms, symptoms that echo the understanding of modern toxicologists. It also was observed that cows and horses should not be pastured near the mines or they would soon become sick and die. Presumably, people living in the communities surrounding the mines would also have suffered environmental exposures from these mining activities.[2]

Paracelsus was the first to attempt a relatively comprehensive description of the diseases of miners and related occupations. He chose a region in Central Europe: this could have been either the Black Forest region in southwestern Germany, Saxony, or Bohemia (now part of the Czech Republic). Others have argued that he described this disease in 1533 among miners in the Tyrol, the mountainous region of Austria.[3] Regardless of the specific region under discussion, *Von der Bergsucht und Anderen Bergkrankheiten* (On the miners' sickness and other miners' diseases), written in 1534, is generally considered to be the first scientific account of the lung disease of any occupational group, making it the first manual in the discipline of what we now call occupational medicine. He named it *mala metallorum*, and it is now believed to be lung cancer caused by arsenic or radioactive dust.[4] His disease descriptions were based on firsthand observations of miners, alchemists, and other processors of metals. He divided his treatise into three books. The first describes the various diseases of miners, the second discusses those of smelter workers and metallurgists, and the third concerns the diseases caused by mercury.

Paracelsus took care to describe the cause, disease progression, symptoms, signs, and therapy of miners' disease. He described the amount of time after exposure before toxic effects manifested. This is the important toxicological concept of latency, and he observed that there were different latencies for different routes of exposure. He wrote, "the air is the body (*corpus*) from which the lung receives its sickness," meaning that miners were exposed to toxins by inhalation.[5] This, in contrast to another route of exposure: ingestion. "For example," he says, "if arsenic is ingested there is a rapid, sudden death; if, however, the body is not taken, but its *spiritus*, the latter makes a year out of an hour, i.e. whatever the body accomplishes in ten hours, the spiritus does in ten years." [6] Here he is recognizing the difference between acute versus chronic forms of poisoning. Eating high doses of arsenic produces instant death, but inhaling its vapor, or *spiritus*, coming off the mineral produces a slowly developing disease with symptoms similar to pulmonary fibrosis, neoplasia, or emphysema.[7]

Paracelsus also attempted to elucidate a mechanism by which exposure produces illness. In smelting, he described how alchemical processes in the furnace separate the air into its constituent elements, which then become harmful to human health. These elements are mercury, which causes disease by coagulating from smoke, roasting sulfur, and salt, which precipitates into the lungs. These are the *tria prima* of Paracelsus, or what we now know as the elements mercury and sulfur and the chemical compound salt, which were his adaptation of the four ancient elements of Aristotle.[8] For Paracelsus, sulfur was a universal component that makes things burn, mercury was the fluid component, and salt the body or solid component. When imbalanced, these elements cannot be properly digested by the lungs and thus form a sort of mucus that induces disease.[9]

Paracelsus did not deny the need for mining. Indeed, he too engaged in various aspects of smelting and metallurgy as part of his interest in alchemy. "We must also have gold and silver," he says,

> also other metals, iron, tin, copper, lead and mercury. If we wish to have these, we must risk both body and life in a struggle with many enemies that oppose us. If we also want to have other things which we are forced to utilize for our healthy life, then there is nothing which doesn't bear our enemy within itself. Because so much lies in the knowledge of natural things which man himself cannot fathom, God has created the physician.[10]

And there we have it. Paracelsus saw himself as the medium between God and the miners. He understood that occupational risks were present but that they were worth the benefits gained from the endeavor.

Georgius Agricola, born in 1494, one year after Paracelsus, was another scholar who studied the diseases of miners. After obtaining his baccalaureate at the University of Leipzig, he studied philosophy, medicine, and natural sciences at universities in Bologna, Venice, and Padua. As a practicing physician, Agricola spent all of his spare time visiting mines and smelters. Like Paracelsus, Agricola was a student of the Peripatetic School of learned

Greeks. However, he had no patience with alchemy or alchemists; he was primarily a mining engineer. In about 1533 he began his major work, *De Re Metallica*, which he completed twenty years later, but it was not published until after his death in 1555.[11]

De Re Metallica was the most comprehensive description of mining and smelting yet, and it would not be superseded for two centuries. Herbert Clark Hoover, a professional mining engineer who eventually became the thirty-first president of the United States, and his wife, Lou Henry Hoover, translated *De Re Metallica* into the English *On the Nature of Metals*. They had met when enrolled in the Geology Department at Stanford University. The text was written in Latin, a language that when Agricola used it had not evolved for about one thousand years, and he had to develop hundreds of new Latin expressions, which deviled his translators, to describe his subject matter.[12] Agricola described the particulars of mining in great detail, and the illustrations in his book are as good as those of any of the Renaissance engravers.

Because he was a physician, the impacts on the health of miners were also part of his observations. In book 1 of *De Re Metallica*, he describes mining as a worthwhile occupation, saying that death by pestilent air and the rotting away of lungs rarely happens, though the consequences are grave.[13] Further on, in book 6, Agricola describes a multitude of mining's effects on health. Miners often stood in water in the mines, and their feet could be eaten away by arsenic and cobalt, which were commonly found in Saxon mines. Dust in dry mines "penetrates into the windpipe and lungs, and produces difficulty in breathing." He adds, "if the dust has corrosive qualities it eats away the lungs and implants consumption in the body; hence in the mines of the Carpathian Mountains women are found who have married seven husbands, all of whom this terrible consumption has carried off to a premature death."[14] While sometimes vivid, Agricola's description of the diseases of miners was nowhere as thoughtful and probing as the treatises of Paracelsus.

• • •

It was not until the seventeenth century that Bernardo Ramazzini wrote more extensively on the diseases in scores of occupations, including mining and metal refining. Ramazzini was born in Carpi, Italy, in 1633, and studied medicine in Parma and Rome. After practicing as a physician in the province of Viterbo, he returned to Carpi to recover from malaria. He achieved some fame administering to a wealthy family in neighboring Modena, and in 1694 he acquired the patronage of Cardinal Rainaldo d'Este, which allowed him to study the workshops of Modena and collect information on the hazards of various occupations.[15]

Besides his direct observation of his patients, Ramazzini drew on information from classical authors such as Hippocrates, Celsus, Galen, Theophrastus, and others as well as sources from the Renaissance, including Paracelsus and Agricola. He had to rely on the writings of learned medical texts for information about some occupations because he was not able to obtain it firsthand. For example, there were no mines near Modena, and apparently Ramazzini never entered one, so he quoted from others who had direct experience. After six years of intensive study, in 1700 he published his major work, *De Morbis Arifficum Diatriba*, which included over sixty chapters describing different occupational settings.[16]

Ramazzini noted that miners frequently suffered from dyspnea, which is difficulty breathing, and phthisis. The definition of phthisis has some ambiguity, because the term had been used to describe diseases caused by dust inhalation, which is pneumoconiosis, and also tuberculosis; however, these two diseases often attacked the same person. Ramazzini also saw high rates of apoplexy (or stroke), swollen feet, loss of teeth, ulcerated gums, pains in the joints, and palsy—symptoms that would be caused by inhaling materials containing metals such as lead or mercury.[17] Ramazzini went much further than Paracelsus in his attention to the prevention of disease among workers. For miners, he recommended fresh air forced in by ventilating machines, protective masks to prevent respiratory-tract disease, and special clothing to protect the skin from exposures.

Ramazzini was especially appalled at the neurological symptoms from which workers in mercury mines suffered. In his chapter on miners, he cites other sources, such as Falloppio and Ettmuller, saying that mercury miners can tolerate their employment for barely three years and that within four months they are afflicted with palsy and vertigo. He also described the afflictions of gilders, who are exposed to poisonous fumes of mercury as it is driven off the gold. Even physicians were exposed to toxic amounts of mercury when they used it as a balm to treat syphilis. Ramazzini attributed the poisoning to dermal exposure; the lowest classes of surgeons, who were employed to apply the ointments repeatedly, were mercury poisoned even when they used gloves.[18]

Ramazzini linked another disease shared by miners, potters, guilders, glass makers, and metal workers to a common underlying cause: lead poisoning. Potters used molten lead in glazes before putting them into the furnace. Ramazzini described the signs of lead poisoning in these potters to be palsied hands, paralysis, lethargy, and wasting syndrome, resulting in a cadaverous face and leaden color.[19] Painters were exposed to lead in paints along with cinnabar and red lead containing lead oxides.[20]

Neither Ramazzini nor Paracelsus described diseases among one important group of mining workers, coal miners, because coal was not a source of fuel in Germany or Italy at the time. Coal mining was an industry based in England and Scotland beginning in the thirteenth century. At that time, they were working above ground, chipping the mineral off outcrops along the banks of rivers. Underground work at any considerable depth had to await the later development of efficient methods of ventilation and hydraulics. In addition, the real incentive for underground coal production would not come until the mass manufacture of iron and the development of the steam engine in the early part of the eighteenth century.[21]

René Laennec (1781–1826) was a French physician best known for inventing the stethoscope. He recognized a disease among coal miners that he called melanosis, in which the lungs were infiltrated with cysts of black matter, what is now popularly called "black lung disease." It was later determined that the inhalation of other minerals, such as silica or asbestos, did similar damage to the lungs, and the generic term "pneumonokoniosis" was introduced in 1866 by the German pathologist Friedrich Albert von Zenker (1825–1898), celebrated for his discovery of trichinosis. Pneumonokoniosis (or, in modern English spelling, pneumoconiosis) means "dusty lung," or fibrosis of the lung, caused by mineral and other types of dusts.[22] Thereafter, the disease among coal miners was dubbed pneumoconiosis anthracosica, or anthracosis for short.

The ancient Greeks had seen pneumoconiosis from exposures to silica in quarry workers. Hippocrates described it, and the Roman naturalist Pliny the Elder claimed that respiratory protection could prevent the disease. Agricola had described this disease in stone cutters, and Ramazzini among miners. The inhalation of silica and other mineral dusts caused chronic inflammation in the lungs, which led to fibrosis and a decrease in lung volume. Sufferers experienced shortness of breath and a chronic cough caused by the accompanying bronchitis. We now call this disease silicosis and are still improving our worker-safety regulations to prevent it.[23]

• • •

By 1875 the lung diseases of coal had, for all practical purposes, ceased to exist as a medical problem in England. This had been achieved by improved ventilation and better sanitary conditions in coal mines, together with shorter working hours.[24] Attention to noxious dust was placed instead on free crystalline silica and the resulting silicosis. In a series of lectures delivered in 1915, E. L. Collis, professor of preventive medicine at the University of Wales, attributed pneumoconiosis among coal miners to silica exposures from sandstone in the mining operations. He believed that the coal itself was not the cause of the pneumoconiosis among coal miners.[25]

But things were not quite so clear-cut. Up until the end of the nineteenth century, the history of the rise and decline of coal miners' pneumoconiosis relied mainly on the recorded experience and opinions of practicing physicians. [26] But in 1896, the invention of the X ray broke the field wide open. William Conrad Roentgen had announced his discovery of a new kind of ray, which he dubbed the "X ray." At first X rays were applied to the diagnosis of fractures and in the detection of radio-opaque foreign bodies such as stones in the bladder. But almost immediately, doctors expanded their use to the investigation of other structures and organs. Thus in 1907, X rays were used to image the lungs of miners with pneumoconiosis in West Australia, and similar lung changes were eventually found among surface workers doing coal screening and trimming at the docks. Thus, coalminer's disease was again found to be an important occupational health problem thanks to the invention of the X ray.[27]

Pneumoconiosis is also caused by asbestos, a disease called asbestosis. This fibrous mineral is composed of silica along with other elements such as magnesium, sodium, and iron. There are three main types of asbestos: chrysotile, which contains magnesium; amosite, containing iron; and crocidolite, which contains both iron and sodium. Although asbestos was reportedly used by the Egyptians, Romans, French, and Chinese, its large-scale use began only in the late nineteenth century, when major deposits were discovered in South Africa and Canada. Because it is remarkably heat resistant, asbestos proved to be extremely valuable as an insulator for boilers, steam pipes, turbines, ovens, kilns, and other high-temperature equipment.

In Quebec, Canada, deposits of chrysotile asbestos were found in 1847, and mining began in the southern town of Thetford some thirty years later. In another thirty years this region of Canada was producing most of the world's asbestos. The proportion later fell as South African, Russian, and Italian mines began operating.[28] Asbestos prospecting and land speculation began on isolated farms in South Africa during the early 1880s, soon after the discovery of diamonds, and during the conquest of African societies by British and Afrikaner colonists. Eventually, prospectors uncovered a rich belt of crocidolite asbestos that extended 240 miles to the Botswana border. The asbestos was marketed to Europe and North America. Chrysotile asbestos deposits were discovered in 1905 near Barberton, and amosite was found in the Pietersburg fields of northeastern Transvaal beginning around 1907.[29]

But soon, the very same British-based companies that were extracting asbestos in South Africa became central players in the identification of asbestos as a cause of lung disease. In 1930, they hosted the International Conference on Silicosis in Johannesburg, where asbestosis was acknowledged as a new occupational disease. Although the precise moment when the industry became aware of the hazardous properties of asbestos is uncertain, in 1928, 1948, and 1959 comprehensive reports were published that clearly showed that asbestos fibers caused asbestosis, lung cancer, and mesothelioma, respectively.[30]

Mining exposures are complicated because silica, metals, and other minerals tend to coexist underground. Just when doctors were getting the hang of using X rays to study lung disease, a new culprit came on the scene with the discovery of gold in the Witwatersrand, South Africa, in 1886. In the course of a few years, examinations of gold miners revealed the deadly dangers of dry drilling in quartz, a mineral similar to silica or silicon dioxide.[31] Beside silicosis, these gold miners were later found to have high rates of lung cancer. The association between gold mining and lung cancer was first described in 1957 in Rhodesia.

The cause of lung cancer among gold miners and smelters is made more complicated by the fact that the Rhodesian gold ore contains high levels of arsenic. Arsenic is present in more than two hundred complex minerals containing silver, lead, copper, nickel, antimony, cobalt, and iron. Therefore, the mining or smelting of any of these ores can cause exposure to arsenic, which has by itself has been found to cause lung cancer as well as skin and bladder cancer. Other types of ore have their own specific risks. Furthermore, mining and smelting of zinc-bearing ores exposed workers to cadmium, which is another cause of lung cancer. Mining of lead chromate (as crocoite) and potassium dichromate (as lopezite) can cause lung cancer from the chromium. Nickel can occur in many minerals, and the mining and smelting of these can lead to cancers of the lung, nasal cavity, and paranasal sinuses.[32]

Silicosis is still with us. Quartz is the second-most-abundant mineral in Earth's continental crust, and sand made of quartz is the most common inland form of sand. This sand is also known as silica. Recently, quartz countertops have become popular for kitchens, and the workers who fabricate them have been found to develop silicosis. When the slabs of reprocessed quartz are cut and drilled, large quantities of silica particles are released into the worker's atmosphere unless streams of water are used to trap the dust.[33] Later, we will describe silicosis associated with the natural gas production method known as hydraulic fracturing. Workers in the production fields have been exposed to the large amounts of sand used in this process.

5 | The Chemical Age

As early as 400 BCE, we have a record of chemical treatment for ailments. The Greek physician Hippocrates at that time recommended a brew made from willow leaves to treat labor pains. About two millennia later, in 1763, an English clergyman by the name of Reverend Edward Stone administered ground-up willow bark to fifty parishioners suffering from rheumatic fever. The active principle from the willow tree was salicylic acid, which was not discovered as a compound until 1828 by Johann Buchner. Salicylic acid would be transformed into aspirin by modern synthetic organic chemistry, and it has become the most common household painkiller in the modern world.

Important changes in thought were needed to allow scientists to develop the field of synthetic organic chemistry. Standing in the way was the ancient belief that a chemical reaction that might take place in the body could not be reproduced in the laboratory. Ironically, it was Paracelsus, the father of toxicology, who was one the historical figures standing in the way.

The archaic school of thought attributed the cause of sickness in man to the failure of Archeus, an occult "vital force," "to perform its function of separating the useful compounds of the body from the poisonous compounds." During the Renaissance, Paracelsus revived the concept of vitalism, which theorized that the functions of living organisms are fueled by a vital principle, the *vis vitalis, spiritus vitalis*, or soul, which is distinct from the physicochemical forces governing the inanimate world. It was believed that the chemical processes taking place in living organisms could not occur without this vital principle. As the laboratory setting lacked this vital principle, organic syntheses in the body could not be replicated in the lab.[1]

However, in a single stroke, this presumption was thrown out the window. The German chemist Friedrich Wöhler in 1828 synthesized the organic chemical urea, which is normally found in urine. By combining cyanic acid and ammonium, the synthesis of an organic compound from two inorganic molecules was achieved for the first time. He announced to his Swedish mentor Berzelius that he had discovered how to make urea in the laboratory without the use of a living kidney. Berzelius was a stern believer in the doctrine of the *vis vitalis*, without which organic compounds could not be generated. Nevertheless, he acknowledged the accomplishment and congratulated Wöhler: "Indeed, the Doctor has mastered the art to make his name immortal."[2] And when Herman Kolbe synthesized acetic acid from carbon disulfide, it was the final blow to vitalism. These discoveries led to the field of synthetic organic chemistry. In 1897, Felix Hoffmann of the Bayer pharmaceutical company synthesized acetyl salicylic acid, the active component in the willow bark: it was later named aspirin.

The age of industrial chemicals began in 1849, when Charles Blachford Mansfield developed a fractional distillation technique that produced pure benzene from coal, which was the first major source of chemicals from fossil fuels. Mansfield was a student of August Wilhelm von Hoffman, the German scientist who had studied coal tar and was recruited by Queen Victoria to London. The extremely flammable benzene is the simplest aromatic compound, having six carbons in a hexagonal structure with three double and three single bonds between the carbons and hydrogen attached to each carbon. Not only is benzene the ultimate solvent for oily substances, it is also the beginning stock of many important synthetic chemicals. Mansfield would tragically become one of the first victims of a laboratory accident: he was severely burned in a laboratory fire and died as a result of his injuries.[3]

Studies of the dye industry provided the first clear evidence that cancer can be caused by synthetic chemicals. A major discovery in organic chemistry occurred when the young English chemist William Henry Perkin, another of Hoffman's students, undertook the synthesis of quinine from benzene. But instead he made a bluish substance with excellent dyeing properties that later became known as aniline purple. Beginning with this beautiful mauve dye, which was much more colorfast than any previously produced, other colors quickly followed. Other components of coal tar soon found their use in the formation of additional dyes and other chemicals. This was the synthesis of the series of important industrial chemicals known as the aniline dyes.[4]

The study of these dyes by surgeons provided important information about how cancer can develop from chemical exposures. In 1895, at the twenty-fourth Congress of the German Association of Surgeons in Berlin, the surgeon Ludwig Rehn gave a speech entitled "Urinary Bladder Tumors Among Fuchsine (Magenta) Workers." During the lecture, he reported finding urinary bladder carcinoma in three out of forty-five aniline workers at the Hoechst paint works. Rehn was an early user of the cystoscope introduced by M. Nitze (1848–1906). Looking directly into the bladders of these workers, who had initially attracted attention because they had blood in their urine, Rehn found two benign tumors and one cancer. According to Robert Case at the Institute of Cancer Research, University of London,[5] Rehn concluded that aniline was the most suspicious of the substances used in the manufacturing process to which these workers had been exposed. This supposition produced the term "aniline tumour of the bladder," which became current in medical textbooks. However, it turned out that aniline was not the culprit but instead other aromatic amines, all of which are nitrogen-added derivatives of benzene.

By 1920 several aromatic amines used in the dye industry were identified as suspects in causing bladder cancer. Wilhelm Hueper, a young, German-trained pathologist, visited DuPont's dye plants in Deepwater, New Jersey, around 1930, and he warned the management that German doctors had found bladder cancer among dye workers in similar employment.

Subsequently, twenty-five cases of bladder cancer were found in these work-ers. DuPont set up a laboratory to investigate and hired Hueper to deter-mine the chemical cause of its workers' bladder tumors. In 1938, Hueper published a paper that reported the production of bladder cancers in dogs with one of the suspected chemicals, beta-naphthylamine. He thereby became the first investigator to demonstrate conclusively a bladder carcin-ogen in a laboratory animal.[6]

Next, T. S. Scott in 1952 described twenty-three cases of benign and malignant tumors of the bladder among 198 workers exposed only to the aromatic amine benzidine from 1935 to 1951 at the Clayton Aniline Com-pany, Ltd., a British dyestuff factory.[7] Robert Case in 1954 reported an anal-ysis of data from twenty-one British factories for the years 1920–1950.[8] Of 298 cases of bladder cancer in a cohort of 4,622 workers, thirty-eight report-edly were exposed to benzidine alone. The expected number of men in an unexposed sample of the general population of 4,622 with mention of blad-der tumors on their death certificates, according to Case, would have been four. This was clearly a dramatic excess of bladder cancer incidence.[9]

• • •

Benzene produced from coal became a major commodity after World War I and subsequently became a ubiquitous chemical used in industry. It was used as a solvent and as a fuel for automobiles. Benzene converted to phe-nol was used in large amounts to make synthetic resin products, such as phenol-formaldehyde resin, and for the production of nylon. Increasing amounts of benzene were used in the manufacture of styrene for synthetic rubber, especially during World War II.[10] Until 1950, benzene was obtained almost exclusively from the products of coal carbonization, either scrubbed from coal gas as "light oil" or distilled from the tar stream. During the 1950s the manufacture of coal-derived benzene reached a peak of almost 200 mil-lion gallons per year, after which it dropped significantly, yielding to petroleum-derived benzene.[11]

Benzene was reported to cause toxic effects on the bone marrow of work-ers who were either making it from coal or using it. Dr. Alice Hamilton, who practiced occupational medicine at Harvard, wrote about some of the earliest-detected health effects of benzene in 1922. She noted that the largest number of cases of benzene poisoning occurred in Germany, although a few cases were reported at Johns Hopkins University before World War I from a factory using benzene to dissolve rubber. What Hamilton was seeing was the destruction of the blood-forming cells in the bone marrow, resulting in greatly decreased counts of circulating red and white blood cells and platelets and in some cases even the complete absence of these components of the blood.[12] Red blood cells are responsible for carrying oxygen, white blood cells fight infections, and platelets are involved in blood clotting. These were serious health effects, and they frequently occurred in a worker immediately after a benzene-poisoning incident.

Two more papers on the occupational effects of benzene were published in 1939. Manfred Bowditch and Hervey Elkins, of the Division of Occupational Hygiene, Massachusetts Department of Labor and Industries, investigated the exposures of workers manufacturing artificial leather shoes, which involved the daily use of rubber cements containing benzene. Francis Hunter of Massachusetts General Hospital described the health records of these same benzene-exposed workers. They found ten cases of fatal poisoning and fifty-seven cases of low red or white blood counts.[13]

It wasn't until 1977 that investigators at National Institute for Occupational Safety and Health, Center for Disease Control, in Cincinnati, Ohio, published a definitive epidemiology study of leukemia in workers producing "Pliofilm" from natural rubber dissolved in benzene. Pliofilm was a moisture-proof membrane made of rubber that was used chiefly for making raincoats and packaging material. Because leukemia was not a rare condition, the rate of leukemia had to be compared to an expected rate in unexposed people. The workplace had been monitored for levels of benzene in the air, and two populations without significant benzene exposures were chosen as control groups for generating the numbers of expected deaths from leukemia in the study population. The first group consisted of the total U.S. white male general population standardized for age and time period over which the study cohort lived. The second group consisted of 1,447 men who had been employed in Ohio at another type of plant. The comparison found that the benzene-exposed Pliofilm workers had about five times the number of leukemia cases compared to the number expected without exposure, and these findings were statistically significant. In another study this same group of investigators evaluated past exposures in both plants in further detail and reported that, although other solvents were used in various areas of both plants, benzene was found to be the only solvent used in the manufacture of rubber hydrochloride (except for chloroform, which was used between 1936 and 1949 in one plant).[14] Furthermore, it would eventually be found that the excess cases of leukemia among the Pliofilm workers were a specific type of leukemia, acute myelogenous leukemia, and this was consistent with the less conclusive earlier findings among workers exposed to benzene in the rotogravure and shoe manufacturing industries in Italy.

• • •

Petroleum eventually emerged as the major source of benzene, overtaking coal production, which further expanded the chemical age. Mocked as a harebrained dreamer, Edwin Drake drilled the first oil well in Pennsylvania in 1859 and changed the history of the world. Soon derricks sprouted throughout Titusville, driven by the unique properties of kerosene, a petroleum product that outshined and would outsell all other sources of light. The first mighty oil gusher, with its towering black jet shooting into the sky, erupted two years after the first well was sunk. John D. Rockefeller Sr., along with his partner Maurice Clark, joined the fray in 1863, and over the next four decades Rockefeller would make his fortune on kerosene and lubricating oils. Gottlieb Daimler fitted bicycles, tricycles, and eventually the motorcar with gasoline engines, and Karl Benz patented an automobile in 1886. Before that, nobody had a use for gasoline, and it was either burned off during the refining process or under cover of night—or allowed to flow into streams and rivers.[15] From these beginnings came the full development of the petroleum industry, which, in addition to kerosene, motor oils, and gasoline, eventually produced petrochemicals such as benzene, toluene, xylene, and ethylbenzene. Benzene was commercially produced from petroleum beginning in 1941. In 1950, petroleum benzene was included in the benzene production statistics for the first time, at ten million gallons.[16]

The chemical era was now exploding, with tens of thousands of new chemicals being produced in the laboratory and thousands of these finding their way into commerce and thus the potential for human exposure. The amount of chemicals that relied upon petroleum and natural gas to manufacture reached 14 billion pounds in 1950, twice as much as the amount of petrochemicals made in 1949. By 1971, the amount of benzene derived from coal sources had shrunk to about 12 percent of U.S. production.[17]

In keeping with the scaling up of petrochemical production, modern toxicology would get legs in the next two decades after 1950 and become an area of study for students in college and graduate school. Before then, toxicology was chiefly concerned with the effects of poisons and how to treat poisonings, or it was studied as the dark twin of pharmacology; while pharmacologists primarily concerned themselves with the beneficial actions of drugs, toxicologists studied their harmful effects. But toxicology expanded to encompass the effects of longer-term chemical exposures occurring in workplaces and the environment. In 1975 the first major textbook of toxicology would be published.[18] A 1979 inventory of chemicals mandated by the Toxic Substances Control Act (TSCA) contained 62,000 chemicals that were reported by manufacturers as being in commercial use at that time.

By this time some of these chemicals had been found to cause cancer. As was described at the beginning of this book, one of these chemicals was vinyl chloride, which had been causally associated with a very rare form of cancer, angiosarcoma of the liver. The demonstration of this relationship had been obvious among the PVC workers. Asbestos was also relatively easy to indict as a cause of a rare lung cancer called mesothelioma among asbestos miners. In the dye industry, chemicals caused the more common bladder cancer, but they had done so in a very high percentage of workers.

It should be noted that the exposures to chemicals in these occupational settings were very high. Dye workers shoveled dyestuff intermediates in bulk and were covered with the dust. As many as half of them developed bladder cancer. Vinyl chloride workers cleaning out reactor vessels were sometimes rendered unconscious by the fumes. The benzene levels in the workers were also high, and there were many cases of bone marrow toxicity.[19] Clearly these workers had been dosed at levels that exceeded the threshold below which the body could detoxify and strike back against the chemicals' cancerous effects.

6 | The Bioassay Boom

H ow should we study the toxic effects of chemicals? Whether an investigator studies humans or animals depends in large part on whether he or she is a physician, epidemiologist, or laboratory scientist. But which approach will give us the most usable information? The answer to this may be discipline specific in some cases, but in general there is value in both and certainly context from a combination of human and animal studies. Nevertheless, throughout history, the topic has been so contentious that scientists would even come to verbal blows over their preferences. Much of medical science has relied on studies of humans; however, the use of animals as models for human anatomy, physiology, and disease has a long history.

The ancient Greek physician Hippocrates (460–370 BCE) relied only on studies of his human patients in order to learn about medical diseases and the effects of toxins. The ancient Greek philosopher Aristotle (384–322 BCE) dissected animals, but his studies contributed more to the developing science of zoology than to any emerging understanding of the workings of the human body. He did not dissect any humans. The first scientific dissections of human cadavers were performed by

Herophilos (325–255 BCE), who received his medical training at the Hippocratean medical school on the island of Cos (Kos) and then moved to Alexandria, Egypt, where he worked with his younger contemporary, the physician Erasistratus. The anatomic and physiologic discoveries of Herophilus were groundbreaking, and because of them he is called the Father of Anatomy.[1]

Galen (129–217 CE) performed dissections of apes, sheep, pigs, and goats, but he did not dissect humans.[2] Galen was able to predict from his anatomical observations some aspects of human anatomy and physiology. For example, he correctly constructed the urinary tract in humans, whereby the urine flowed from the kidneys through the ureters and into the bladder.[3] Human dissection was considered blasphemous during the Middle Ages. That particularly strict era of Christianity paralyzed rational thought and taught that physicians could do no better than to repeat the works of Aristotle, Galen, and other eminent figures of the past.[4] It wasn't until later that William Harvey (1567–1657) would correct some of the mistakes made by Galen. Trained as a doctor of medicine at Cambridge, Harvey based his conclusions largely on new observations in animals and not by examining human cadavers.[5]

Human dissections were permitted on a limited basis beginning in the thirteenth century, and the Italian Mondino dei Luzzi wrote a treatise on anatomy in 1316 CE; however, his findings were little different than those already established. Subsequently, studies by many other anatomists, including the artist Leonardo da Vinci, led to the definitive work by Andreas Vesalius.[6] As we saw in earlier chapters, medicine was transformed by the studies of Paracelsus in the sixteenth century and Ramazzini in the eighteenth century, studies involving the examinations of humans.

The pendulum then swings back to animal studies. Louis Pasteur (1822–1895) was one of the first scientists to use animals experimentally to fight disease. Pasteur was one of the developers of the germ theory of infectious disease, which stated that foreign agents—what we would now recognize as viruses or bacteria—can invade the body and cause infection. Pasteur suspected that if such an agent were introduced in a weakened form, it would arouse the immune system and enable it to prevent or fend off infection. After studying rabies, he found that the dried spinal cord of infected rabbits contained an almost nonvirulent microbe that he could inject into dogs, which would prevent them from developing rabies when he exposed them to the virus. A dramatic test of his rabies vaccine came when he was asked to treat a child who had been bitten fifteen times by a dog who was thought to be rabid. Based on what he had learned from his animal studies, he gave the child greater and greater doses of his vaccine over a fourteen-day period, and the boy remained well. After a second similar success, Pasteur had demonstrated the value of experimental animal studies in fighting human disease, and in the ensuing national enthusiasm the Institut Pasteur was established in 1888.[7]

Robert Koch (1843–1910) isolated the anthrax bacillus that Pasteur used to develop his vaccine. He also isolated the causative agents for tuberculosis, diphtheria, typhoid, pneumonia, gonorrhea, meningitis, leprosy, plague, tetanus, syphilis, whooping cough, and others. Moreover, in a tour de force of the scientific method, Koch developed a set of principles required to establish an organism as a cause of disease, which came to be known as Koch's Postulates. One of these involved testing the purported pathogen in animals: "That the disease could be reproduced in experimental animals through a pure culture removed by numerous generations from the organism initially isolated."[8]

Animals continued to be important in the study of many aspects of human disease. In toxicology they were essential because only in animals could toxins be administered intentionally, so that their effects could be observed. Eventually, research on animals would become important to our understanding of how chemicals can cause chronic human diseases including cancer, but this would be much less straightforward than isolating infectious agents or describing the deadly effects of venoms.

• • •

Three striking events led to the widespread use of animal bioassays in toxicology: the poisonings caused by a new formulation of sulfanilamide, the thalidomide birth defect disaster, and the publicity surrounding the congressional Delaney Committee hearings. The first "sulfa" antibiotic drug sulfanilamide was introduced in the 1930s to treat bacterial infections. It had dramatic curative effects when administered in tablet and powder form. But the pharmaceutical company S. E. Massengill Co. of Bristol, Tennessee, wanted to offer a liquid form for use in adults and children; they experimented and found that sulfanilamide would dissolve well in the newly developed solvent diethylene glycol. The company compounded a quantity of what they called Elixir Sulfanilamide and sent hundreds of shipments out all over the country.[9]

Soon horrific reports of sudden deaths started coming in. One physician wrote in 1937:

> But to realize that six human beings, all of them my patients, one of them my best friend, are dead because they took medicine that I prescribed for them innocently, and to realize that that medicine which I had used for years in such cases suddenly had become a deadly poison in its newest and most modern form, as recommended by a great and reputable pharmaceutical firm in Tennessee: well, that realization has given me such days and nights of mental and spiritual agony as I did not believe a human being could undergo and survive. I have known hours when death for me would be a welcome relief from this agony.[10]

In a letter to President Franklin D. Roosevelt, a woman described the death of her child:

> The first time I ever had occasion to call in a doctor for [Joan] and she was given Elixir of Sulfanilamide. All that is left to us is the caring for her little grave. Even the memory of her is mixed with sorrow for we can see her little body tossing to and fro and hear that little voice screaming with pain and it seems as though it would drive me insane. . . . It is my plea that you will take steps to prevent such sales of drugs that will take little lives and leave such suffering behind and such a bleak outlook on the future as I have tonight.[11]

The law at that time did not require that safety studies be done on new drugs, and the new formulation of sulfanilamide had not been tested for toxicity. Once alerted, the pharmaceutical company did animal tests on their Elixir

Sulfanilamide and found that the diethylene glycol used to dissolve the sulfanilamide drug was a deadly poison.

The Elixir Sulfanilamide fiasco hastened the enactment of the 1938 Federal Food, Drug, and Cosmetic Act, which required the use of animal testing for new pharmaceuticals to ensure that they wouldn't be toxic for humans. Twenty-five years later, this new system of drug control would save the United States from an even greater drug tragedy—the thalidomide disaster that occurred in Germany and England.[12]

The first paper describing the pharmacological actions of thalidomide was published in 1956, and it was first marketed in Germany the next year. Thalidomide was shown experimentally to be an effective sedative-hypnotic drug acting differently from the barbiturates. Thalidomide was thought to be virtually nontoxic because it did not cause incoordination, respiratory depression, or narcosis.[13] In Germany, untested drugs were much easier to get to the market, and thalidomide was sold as a nonprescription, over-the-counter product. In England at that time, there was similarly no requirement for the confirmation of a drug's effectiveness or safety.[14]

Thalidomide's alleged lack of toxicity would soon come tragically into question. At a meeting in Germany in December 1959, a pediatrician named Weidenbach presented the history of a deformed baby born the previous year. It was thought that the cause was hereditary. However, nine months later at the German Society of Pediatrics, two other physicians described two more infants with similar malformations. Apparently, nothing much was made of these findings until Weidenbach found more cases and published a scientific paper describing thirteen cases of a syndrome called "phocomelia," involving a shortening or absence of the extremities along with other dramatic birth defects. No such cases had been reported in the previous decade in Germany. This was a genuine birth defect cluster.

Only two months later at another meeting in Germany, Peiffer and Kosenow reported on thirty-four infants with long-bone defects in a twenty-two-month period at the Children's Hospital in Munster. Dr. Widukind Lenz, of Hamburg, who attended the meeting, wondered if these could be related to drugs, and, sure enough, his investigation revealed that many of the mothers had taken the new sedative thalidomide for morning sickness.[15] By November 16, 1961, he felt sufficiently certain to warn the pharmaceutical manufacturer Chemie Gruenenthal of this problem. At a medical conference on November 18, Dr. Lenz discussed the possible role of thalidomide in the production of these anomalies. He had noted that a considerable number of the mothers of infants so affected had been taking this drug during the second month of pregnancy, the period of active development of the fetal structures in which malformations were evident at birth.[16]

It turned out that in 1961 alone, 477 cases of these previously rare types of congenital malformations of newborn infants had been observed in West Germany. The tragedy cascaded, with reports of similar abnormalities coming in from a number of other countries, notably East Germany, Belgium, Switzerland, Sweden, Australia, England, and Scotland. Eventually health workers identified about eight thousand deformed infants in some thirty countries whose mothers had taken thalidomide for morning sickness.[17]

The United States was mostly spared from the disaster by the efforts of a Food and Drug Administration medical officer named Dr. Frances O. Kelsey, who had squelched approval of the drug. Kelsey took her stand against thalidomide during her first month of employment at the Food and Drug Administration: thalidomide was her first assignment. The task was supposed to be a straightforward review of a sleeping pill already widely used in Europe, but Kelsey was concerned by some data from the pharmaceutical company suggesting dangerous neurological side effects in patients who took the drug repeatedly. She continued to withhold approval, and the William S. Merrell Company of Cincinnati, the U.S. manufacturer, did everything it could to get around her concerns.[18]

Although Kelsey was new to the FDA, she had the proper training to pass judgment on the application. She had an MD and a PhD in pharmacology from the University of Chicago. "I must say I was shocked at the caliber of work that would go into these applications in support of safety," Kelsey later recollected to the London *Times*. While the 1938 FDA law required some testing of new drugs, it was written in such a way as to permit the approval of a drug unless there was good reason to stop its approval—in effect, an "innocent until proven guilty" policy.[19] In humans, there were no long-term preapproval studies for thalidomide, and the animal data was sketchy. There were some reports of the drug causing an effect on the sensory nerves, and there was evidence of more widespread effects on the nervous system.[20]

Kelsey used these findings as the basis for holding up the drug's approval in the United States. As a result of her efforts, there were only seventeen malformed babies in the United States, seven of whose mothers had obtained the drug abroad and ten who had been medicated under an exemption for investigation of the drug.[21] On August 7, 1962, Frances Kelsey received the Distinguished Federal Civilian Service Award from President John F. Kennedy.[22]

• • •

The thalidomide tragedy in Europe focused public attention on pending U.S. legislation to strengthen the Federal Food, Drug, and Cosmetic Act, and meanwhile Congress passed the Kefauver-Harris Amendments. Although previous governmental recommendations included animal testing during the reproductive cycle of animals, a Commission of Drug Safety eventually caused the FDA to require additional animal testing preceding large-scale human clinical trials. This included a specific test for birth defects, and women of childbearing potential were not allowed to take investigational drugs until these animal tests were completed.[23]

The Drug Amendments of 1962 also tightened control over the approval of new drugs in other ways. The FDA set up a new branch to test and regulate new drugs, and Kelsey was put in charge. Before that, drugs were being sold for all sorts of unsubstantiated claims of medical effects. In the Drug Amendments of 1962, it was recognized that a drug had to be proven to be effective for the prescribed condition as well as safe before it could be marketed, a milestone in medical history. Consequently, in the years after 1962 thousands of prescription drugs were taken off the U.S. market because they lacked evidence of safety and/or effectiveness or had to have their labeling changed to reflect the known medical facts. In her forty-five-year career with the FDA, Kelsey continued to strengthen the rules. She eventually became director of the agency's Office of Scientific Investigations.[24]

The FDA had earlier initiated policies requiring animal testing for food additives. The U.S. government had been in the business of protecting the public from adulterated food and drugs since the early twentieth century, but it was not until the 1950s that it became concerned about potential carcinogens in food, drugs, and the general environment. In 1949, FDA Commissioner Paul B. Dunbar convinced Rep. Frank B. Keefe of Wisconsin to introduce a resolution to investigate chemicals in food.

The House Select Committee Investigating the Use of Chemicals in Food and Cosmetics, led by the New York Democrat James Delaney, was convened in 1950 to hear testimony on carcinogens in the food supply. These proceedings, which would be known as the Delaney Committee hearings, laid the foundation for expanded animal bioassays designed to identify carcinogens. As a result of the hearings it was soon clear from the information presented to the committee that DDT and other, more powerful synthetic pesticides, such as parathion, chlordane, and heptachlor, were toxic. In 1954, Congress passed the Miller Pesticide Amendment to the Food, Drug, and Cosmetics Act, which set safe limits, based on human exposure, for pesticide residues on raw agricultural commodities.

Also of particular concern was contamination from the powerful estrogen diethylstilbestrol, or DES. Pellets of DES were injected into the necks of chickens to fatten them up. It was suspected that estrogens increased the risk of breast cancer, and the meat of these chickens contained DES residues. The Delaney Committee hearings on DES eventually led to the 1958 Food Additives Amendment to the Food, Drug, and Cosmetics Act, which prohibited the use of food additives found to be carcinogenic in either laboratory animals or humans. Although the new amendment put an end to the use of DES in chickens, it would remain on the market as a drug given to humans to prevent miscarriage until the 1980s, when overwhelming evidence revealed that it caused reproductive cancer and abnormalities in the children of women who had taken it.

The Delaney Committee eventually produced three amendments to the U.S. Food and Drug Law and defined how the regulatory process would work going forward. But by the time the Delaney amendments were passed, thousands of drugs and chemicals were already on the market. These amendments required the food industry to use animal testing to prove the safety of food, including food and color additives and pesticide residues. The FDA specified protocols for developing the animal data necessary to evaluate the safety of food and color additives beginning in 1955, and by 1973 the requirements for animal testing for investigating and marketing new human drugs included an eighteen-month rat study and a twelve-month study in dogs or monkeys. Eventually the Pharmaceutical Manufacturer's Association decided that the testing used for a New Drug Application should include eighteen- to twenty-four-month chronic studies in rats and mice.[25]

• • •

John Weisburger became a U.S. Public Health Service research fellow with the National Cancer Institute (NCI) in 1950, the same year that the Delaney Committee began its public hearings that led to the bioassay boom. Weisburger was named head of the Carcinogen Screening Section at the NCI and, later, director of the Bioassay Carcinogenesis Programs, where he along with his wife, Elizabeth, began studying the mechanism by which the aromatic amine 2-acetylaminofluorene (AAF) causes cancer.

When Weisburger started the carcinogen testing program in 1961, studies were quite focused on comparing the carcinogenic potential of aromatic amines that were chemically related to AAF with those that had already been proven to cause bladder cancer in the German dye workers studied by Rehn in 1895. John and Elizabeth Weisburger developed the protocols for animal tests, and the first priority for using bioassays was given to testing chemicals that would contribute to the understanding of how cancer develops.

Soon the growing demands of government regulation required the testing of hundreds of chemicals, which overwhelmed the fledging animal testing industry.[26] The Federal Insecticide, Fungicide, and Rodenticide Act of 1947, administered by the Department of Agriculture, was amended in 1972 to require animal testing of pesticides for carcinogenicity and was brought under the newly created EPA. There were six hundred active ingredients that needed testing in order to comply with the new regulations.[27]

The design of the animal bioassay, as it is currently practiced, was standardized in the 1970s. It usually calls for two separate sets of tests, one in mice and one in rats. Each set uses three different doses of a chemical, with one hundred animals per dose group and another hundred animals designated a control group. The highest dose is that which is tolerated by the animals during their lifetime dosing without significantly increasing their mortality from toxic effects. This is called the Maximum Tolerated Dose (MTD), and it is extrapolated from additional, shorter-term tests preceding the main, lifetime tests. Including all phases of testing from LD50 to two-year studies, each bioassay—the testing of a single chemical—usually employs well over a thousand animals and takes several years to complete.[28]

The purpose of the bioassay is to show toxic effects, and the dose is pushed up to cause these effects. In the years before bioassays were used to test for carcinogenicity, the primary purpose was to find the toxic effects caused in specific organs; this was especially important for pharmaceuticals and food additives. Animal tests were and are used to find out which organs might be the most susceptible to the toxic effects of a new drug so that when clinical trials are planned, more extensive physical examinations and laboratory tests can be directed at those organs in humans. In the case of testing in whole animals, toxicity may also be attributable to events outside the cell, such as alterations in blood pressure, heart rate, respiration, and so forth.

But why should animal testing use such high doses for detecting cancer? Why not test at the levels of typical human exposures? The answer involves the rate of detection of toxic effects, which is quite low for each exposed animal or human. After specifying the level of statistical confidence desired, one can calculate the minimum number of animals required to detect a toxic effect. For detection of an effect that happens in 5 percent of animals in a statistically reliable way, fifty-eight test animals are required; for effects in 1 percent of animals, about three hundred are needed.[29] One can extrapolate this out to three thousand or thirty thousand animals for frequencies of events of 0.1 percent and 0.01 percent, respectively. Clearly, we do not have very good sensitivity for most cancers in a bioassay test if we only use fifty animals per group. Regulatory toxicologists came up with a clever way of overcoming this obstacle by increasing the dose given to the animals, because dosage effects usually increase linearly. So instead of giving thirty thousand animals a low dose that might approximate the usual human exposure, one gives about five hundred times the human exposure, and voila, we only need about fifty animals per dose. The only problem with this scheme is that we may be poisoning the animals at this dose and causing effects that would never occur at human exposure levels. And chronically poisoning an animal in a particular organ may lead to cancers in that organ that would not occur in the absence of toxicity.

Toxic effects usually don't occur at low exposure levels because of what are called homeostatic mechanisms. These can be physiological mechanisms that counter the effects of a chemical, for example, a stimulus to raise the blood pressure when a toxin causes a drop in blood pressure. Another example is the regrowth of cells damaged by a toxic insult. But when an animal is given a chemical in an amount five hundred times that of a potential human exposure, we overload the homeostatic mechanisms and overwhelm the body's ability to prevent or repair the toxic insult. So, the "maximum tolerated dose" toxic effects in rat and mice might never occur in humans at the usual therapeutic doses of a medicine or occupational exposures to chemicals, let alone environmental exposures. Once animal test results are available, the difficult and relevant task is to find out whether humans have the same proclivity as animals, or lack thereof, for the demonstrated animal effects at human doses.

• • •

The National Toxicology Program (NTP) was established in 1978 in the Department of Health, Education, and Welfare, and most animal carcinogenicity testing was transferred there from the NCI in 1981. The NTP was given the purpose of performing the animal testing not only for the NCI but also the FDA, the National Institute of Occupational Safety and Health (NIOSH), and the National Institute of Environmental Health Sciences (NIEHS). The NTP determined that of the approximately 5 million chemicals that they reviewed, humans were exposed to about 53,000. Of these, the NTP estimated that humans could be highly exposed to approximately one thousand chemicals, and these were the ones that should be tested in their program.[30]

But the extent of required animal testing for these thousand chemicals exceeded the in-house capabilities of the chemical manufacturers, pharmaceutical companies, and the NTP. It became necessary to expand greatly the use of contractor laboratories to provide the necessary testing services, and this became a big business. One of these contract laboratories was Industrial Bio-Test Laboratories, founded by Dr. Joseph Calandra. Dr. Calandra resigned the presidency of Industrial Bio-Test in 1977 after the FDA found that data developed for the antiarthritis drug Naprosyn was flawed. The FDA found that the study records were inadequate, that tumors in some animals had not been reported, and that examinations of some animals after their deaths were unreliable because of their advanced stage of decomposition.[31] According to grand jury testimony disclosed by Industrial Bio-Test's attorneys, many of the test animals were housed in a waterlogged room known as "the Swamp." "Dead rats and mice," technicians later told federal investigators, "decomposed so rapidly in the Swamp that their bodies oozed through wire cage bottoms and lay in purple puddles on the dropping trays." Mortality was so high that it was impossible in some cases to assess the effect of the chemical under test.[32]

The EPA had approved more than one hundred chemicals based on Industrial Bio-Test's research. They had to declare the tests invalid and advise the manufacturers to retest their products.[33] Three of Industrial Bio-Test's officials were convicted of falsifying drug tests and received prison sentences of up to thirty years, and Industrial Bio-Test closed its doors in 1978.[34] All of these problems only exacerbated the problems of testing chemicals by U.S. governmental agencies. The deadline set for 1976 to test all of the 1,400 ingredients used in 35,000 pesticides was a decade behind schedule by 1977. The Senate committee overseeing this effort said that the system was in a "state of chaos."[35]

The Industrial Bio-Test fiasco prompted Congress to convene the Kennedy Hearings to investigate laboratory practices by the NTP's other contractors. The inspection of studies and facilities at various testing companies revealed instances of inadequate planning, incompetent execution of studies, and insufficient documentation of methods and results. There were even cases of fraud in which animals that had died during a study were replaced with new ones (which had not been treated appropriately with the test compound) without documentation. Gross necropsy observations were deleted because the histopathologist, who was supposed to examine the same tissues microscopically, received no specimens of these lesions. As a result of these hearings the FDA published the Proposed Regulations on Good Laboratory Practices (GLP) in 1976, with the establishment of the Final Rule in June 1979, which ensured that such tests would be done properly from then on. Eventually the EPA would adopt the same testing procedures.[36]

The necessities of the standardized study design and the greater workload involved in GLP requirements force-fed a ballooning animal bioassay industry. Governments worldwide regulate drugs, chemicals, and products such as petroleum involving over a trillion dollars of yearly sales. In 2016, it was estimated that there were 12 million animals in use for studies in the United States.[37] The results of these studies are crucial for the pharmaceutical industry, which boasts $200 billion in yearly sales in the United States alone. The cost for testing in Europe, which uses about one million animals, was estimated to be €620 million per year. Worldwide, the cost is demonstrably in the billions of dollars.[38]

Chemicals have now been tested for cancer in the United States in over a thousand GLP animal bioassays involving rats and mice. As will be discussed in subsequent chapters, the challenge has been to interpret all of this data properly, especially for its relevance to humans. Although the animal bioassay is still performed routinely for pharmaceuticals, the numbers of bioassays for other types of chemicals has become less common. For example, the NTP issued about three hundred technical reports for cancer bioassays from 1976 to 1985, about 140 during the next ten years, and about the same amount over the next twenty years. On the other hand, a search of PubMed citations shows that since 1976, the number of epidemiology studies reported has been increasing every ten years. Consequently, there appears to be a growing amount of human data being generated compared to animal studies, which may be an indication that there is a belief that human studies are more reliable in predicting disease from chemicals.

How Do We Study Toxicology, and What Have We Learned?

It was all very well to say "Drink me," but the wise little Alice was not going to do that *in a hurry; "No, I'll look first" she said, "and see whether it's marked 'poison' or not" . . . and she had never forgotten that if you drink very much from a bottle marked "poison," it is almost certain to disagree with you, sooner or later.*
—Lewis Carroll, *Alice in Wonderland*

This group of chapters begins with two examples where toxicology has helped us understand clinical and ecological observations. The first is childhood lead poisoning and the investigation by pediatricians and toxicologists to discover the sources of lead and determine lead's effects on the developing mind. The second example is Rachel Carson's study of the indiscriminate use of pesticides and its effect on wildlife. Next, we turn to the study of cancer and the chemical causes of cancer. Through research in laboratories, it became clear that there are several types of mechanisms involved in the chemical causes of tumor in animals and humans; all cancers are not created equal. Genotoxicity is very important for certain chemicals.

And toxicity that produced mutations through secondary mechanisms involving the role of oxygen attacks on DNA is also involved. Consequently, the field of toxicology made important contributions to the understanding of how chemicals can cause cancer.

However, the most important discovery of cancer caused by chemicals came from the study of lung cancer in humans. Through a combination of epidemiology and toxicology studies, Ernst Wynder, Richard Doll, Austin Bradford Hill, and their associates found tobacco to be the single most important preventable cause of cancer. Other major cancer causes were found to be diet, including obesity; infection; and occupational and environmental exposures. The final chapter in this group explores the question of how much cancer is preventable and how much is caused by the unavoidable, endogenous cellular processes taking place within us as we age.

7 | Lead

A Heavy Metal Weighing Down the Brain

You might wonder why a substance as toxic as lead was ever used in paint. Painters mixed white lead with oil and mineral spirits to make paint that contained about 50 percent lead. They were convinced that lead added durability to the paint that they applied to the wooden exterior of houses. Lead was also in such artists' oil paints as flake white and lead chromate. And lead was prized as a highly useful substance for water pipes and many other applications.

Lead poisoning, which was originally described in miners by the Greek physician Hippocrates and explored by others, including Paracelsus and Ramazzini and throughout the Enlightenment, came to prominence in the United States at the beginning of the twentieth century. That was when Alice Hamilton began studying lead poisoning in her work for the Illinois Occupational Disease Commission.[1] Hamilton discovered cases of lead poisoning among workers producing the white lead pigment used in paint. Hamilton's important studies of lead and other causes of occupational disease will be described further in a later chapter.

In 1974 the Department of Housing and Urban Development (HUD) became involved in the lead paint issue. They had a big problem because they were the largest landlord in the United States and consequently controlled plenty of housing with lead paint. The program was manned by two stimulating and thought-provoking HUD employees named David Engel and Gene Gray. Gene was the quiet type and served as a sounding board for David; David was a bundle of energy and had been saddled with this huge problem. His expertise was in the development of affordable housing. But he tried hard to solve HUD's lead difficulties.

David needed to find out how much lead should be allowed in commercial paint products. When hiring me as a HUD consultant, he asked me to solve what appeared to be a simple question: how much lead paint at the existing standard could a child eat without getting lead poisoning? It was not until the late 1950s that the U.S. paint industry began voluntarily restricting the amount of lead in commercial paints to below 1 percent because of growing concerns about its toxicity.[2] In 1974 the allowable amount was half a percentage point. This was a hundred times less than the amount of lead in paint in previous decades.

It was clear that the traditional old paint could cause lead poisoning, but was it safe for kids to eat paint at the new standard? Using the 0.5 percent allowable amount of lead in paint and an assumption of a child eating a good-sized paint chip, I calculated the resulting blood lead level. If all the lead stayed in the blood—in equilibrium with the brain, where the damage is done—I found that the brain's lead level could become dangerously high quickly and that the child could swiftly be lead poisoned.

Next, I went to the National Library of Medicine at the National Institutes of Health, where I had been a postdoctoral fellow a few years earlier. Perusing the medical literature, I learned that the biggest complicating factor in my calculations was that large amounts of lead can be stored in the bones, where it hovers in equilibrium with the lead in the blood. So if a child eats small amounts of lead on a daily basis, the lead level may at first be low but then insidiously reach much higher levels. I also learned that in Baltimore, the city health department found lead poisoning especially common in children who exhibited a condition called pica, which is the compulsive eating of nonfood substances.[3] These pica children would eat lead paint on a daily basis. So my answer was that lead at the existing standard in paint could cause lead poisoning and needed to be lower.

David always looked for innovative solutions to the lead problem. For example, he was fascinated by a manufacturer in Brooklyn who had the idea of developing a bad-tasting paint containing denatonium benzoate, the bitterest substance known. The idea was to include it in the paint so that children would be repulsed from eating any paint chips. Even covering the old paint with the bitter paint would keep children away. This approach was attractive because there were many difficulties with removing lead paint from houses. Either sanding or burning the paint off could cause exposures to workers or even to the children they were trying to protect. Often the poor condition of the underlying wood or plaster meant that it had to be replaced once the paint was removed, making the remediation very expensive.

To develop this paint, we combined toxicity studies in rats with a testing program for children to determine whether such a product would be safe and effective. Trying to find a way to see if children would be repulsed by the paint required testing it in toddlers. I found a university pediatrician who had the perfect place to test this strategy. Wooden blocks were painted with the paint, and the kids were observed to see if they would be repulsed when they tried to put it in their mouths. It worked. Later, this paint was introduced for sale under the product name ChildGuard. Today, ChildGuard has been approved for use in all fifty states.

David also helped advance the technology of lead paint detection by funding research on a handheld X-ray fluorescence (XRF) detector. Like many elements, lead emits X rays at a characteristic frequency when exposed to high-energy radiation. The intensity of the rays can be measured and correlated to the amount of lead. Thus, with a handheld XRF device an inspector can detect the presence of lead paint in real time, rather than having to send samples off to a laboratory for testing.

Lead detection and intervention programs were up and running in dozens of cities in 1974, but they met with difficulties in bringing down the blood lead levels of affected children. Removing the lead paint source from the child's environment should cause a decrease in the blood lead level. But when lead is removed from the blood—through the urine, for example—it is quickly replaced from the large store of lead contained in the bones. Thus, despite some lead leaving the body when the source is removed, the bone reservoir maintains the high level of lead in the blood.

At first it was thought that the programs were not effective at eliminating the sources of lead paint from the housing. But even with the best interventions, the children's blood lead levels often did not decline. So what was happening, and what were these programs doing wrong? Part of this could be attributed to the large bone reservoir for lead, causing the slow elimination from the body. But even in cases where the lead paint had been completely eliminated for a year, lead levels in the blood remained stubbornly high or even went up. Vernon Houk soon provided an answer to this puzzle in what was one of the great "a-ha" moments in toxicology. He announced to HUD that they had missed the other major source of lead in the environment of the typical child: gasoline.

While HUD was trying to figure out how much lead in paint could be tolerated and how to detect it, Dr. Vernon Houk at the Centers for Disease Control (CDC) tried to improve lead-based-paint poisoning prevention programs across the United States. Houk was among the first health officials in the country to recognize that lead in gasoline wound up in the air, dust, and soil and that these exposures constituted an additional major ongoing source of lead exposure for children. This helped solve the mystery of why some children's lead levels did not go down as expected, especially in urban settings. That is, children who were being treated for lead-based-paint poisoning were at the same time receiving a regular dosage of lead from their environment, from leaded gasoline. We were seeing two sources of a toxic exposure causing the same disease.

• • •

Lead was added to gasoline beginning in 1923, in the form of tetraethyl lead, which boosted the fuel's octane rating, a measure of combustion performance. Discovered in 1852 by Karl Jacob Lowig, a German chemist, tetraethyl lead long had languished in obscurity, a rare and poisonous compound with no apparent use. Then General Motors gave a mechanical engineer named Thomas Midgley Jr. the task of developing a low-cost fuel as part of GM's push to try to overtake Ford's Model T. General Motors cars were poky and would knock going uphill because of premature detonation of the fuel in the high-compression cylinder. When this happened, only about one-twentieth of the fuel's potential energy was released. After three years and many disappointments, Midgley rediscovered tetraethyl lead and mixed it with gasoline. This new fuel exceeded all expectations for engine performance when just 0.05 percent of tetraethyl lead was added to the gasoline.[4]

Almost from the start, health authorities were concerned about the potential of leaded gasoline to cause poisoning. After all, lead's toxic effects had been known for two thousand years. Even Midgley was concerned about the possible effects of breathing the automobile exhaust, especially on heavily traveled roads and in tunnels. Unlike lead oxide, the inorganic form of lead found in automobile exhaust, tetraethyl lead is an organometallic compound and thus can be easily absorbed directly through the skin. In fact, Midgley and his assistant developed acute lead poisoning in the course of their laboratory work. And in 1924, two workers at a tetraethyl plant in Dayton, Ohio, died of lead poisoning, and sixty others became ill. Even as lead poisoning quickly became an occupational health issue after the rediscovery of tetraethyl lead, the effects on the general public of automobile exhaust were ignored. But although health studies of lead exposure from auto exhaust were planned, none ever materialized.[5]

Perhaps because of this lack of public oversight there was no stopping the marketing of tetraethyl lead. The first stratagem in the public relations arena was to drop the word "lead" from tetraethyl lead and shorten it to "ethyl," the product of the Ethyl Corporation. Next, Dr. Robert A. Kehoe, a young pathologist, was enlisted by the corporation to debunk its health hazards. He stated that the additive posed no danger for the general public and that a few simple safety regulations would eliminate the danger to workers in the factories that produced it.[6] Despite these claims, it wasn't long before an independent study found that the amount of lead in street dust had increased by 50 percent in the decade spanning 1924 to 1933. However, these findings were largely ignored.[7] By the 1960s, lead contamination in the air had become so ubiquitous in the United States that it was difficult to study lead in the laboratory.

In 1953, a geochemist named Clair Patterson of the California Institute of Technology had used lead isotopes to make the first reliable estimate of Earth's age: around 4.5 billion years. In the process, he had become proficient at detecting lead at extremely low levels. However, the biggest obstacle to doing his research was the ubiquitous lead contamination in his laboratory. He used his meticulous analytical methods to reveal the pervasive presence of lead in the planet's natural environment, even in snow and seawater—a circumstance that Patterson attributed to industrial sources.[8] In 1965, he published a study of the Greenland ice sheet showing that lead from gasoline had caused an alarming increase of lead in the air all over the world. Three days after his article was published in *Nature*, corporate representatives of the lead industry visited him in his office and attempted to buy him off. He refused their offer and soon afterward lost both the governmental and industrial funding for his research. A member of Caltech's board who was an executive of a petroleum company that sold gasoline containing tetraethyl lead pressured Lee Dubridge, the president of Caltech, to curb Patterson's ongoing research. To his credit, Dubridge refused to interfere with the geochemist's work.[9]

In 1975, spurred by Vernon Houk, who had warned us about this contribution to the lead levels in some children, the U.S. government finally mandated that only unleaded gasoline could be used in new cars. Additional regulations phased out leaded gasoline entirely by 1986. The Second National Health and Nutrition Examination Survey conducted from 1976 to 1980 showed a decrease of about 40 percent in blood lead levels, corresponding with the decrease in the amount of leaded gasoline used and the abatement of lead in housing.[10]

Despite this progress, Dr. Julian Chisolm at Johns Hopkins University in Baltimore estimated in 1979 that there were still forty thousand children per year being identified with elevated lead levels in the United States.[11] At the same time, neurological injuries in children were being found at lower and lower exposure levels. The health impairments originally recognized in 1890 by Turner and Gibson in Australia as requiring hospitalization had stemmed from lead's acute toxic effects on the brain and gastrointestinal tract. They included anemia, kidney damage, colic (severe "stomach ache"), muscle weakness, and brain damage, which ultimately could prove fatal.[12] However, research had shown that lower levels of exposure could result in less severe but still detrimental effects on the nervous system.[13]

Thus, from the 1960s through the 1990s the level of lead considered to be acceptable in children decreased dramatically. In the early 1960s the upper limit of "normal" lead in the blood was 80 micrograms of lead per 100 deciliters of blood. This is approximately one part per million but is usually abbreviated as 80µg/dL. In children the acceptable level was considered to be 60µg/dL, but by the 1970s this had been reduced, and the rule of thumb was that a child's level should be no greater than 40µg/dL. Then in 1975, 1985, and again in 1991, these levels were reduced to 30, 20 and then 10µg/dL, respectively. Methods for measuring the effects of lead on the developing nervous system in children had become increasingly more sensitive over this sixteen-year period.[14] By the end of the 1980s, studies had confirmed that blood lead was associated with clinical attention deficit/hyperactivity disorder (ADHD) at lead levels of 10µg/dL or less.[15] Later studies found that even blood lead levels as low as 1 to 10µg/dL were linked to a lower child intelligence quotient, weaker executive cognitive abilities, and ADHD in community surveys.[16]

There is a growing body of research suggesting that lead exposure causes not only juvenile learning deficits but also an array of antisocial and even criminal behaviors later in life.[17] The impetus for this research was the hypothesis that the rise in crime between 1960 and 1990 was driven in part by increased exposure of young children to environmental lead from the 1940s to the 1980s. Crime had surged twenty years after the first children exposed to lead from gasoline and paint entered young adulthood, the life stage at which a majority of crime is committed. According to this hypothesis, the decline in crime after 1990 was a result of efforts to protect children from the most severe environmental lead exposures, notably the removal of lead from gasoline in the United States from 1975 to 1985.[18]

Several studies have provided substantiation to this hypothesis that lead levels in childhood and crime are associated. For example, prenatal (and childhood) blood lead concentrations predicted arrests during adulthood in a cohort of more than two hundred subjects in the Cincinnati Lead Study. Another study analyzing the association between elevated blood lead levels and crime across census tracts in St. Louis, Missouri, found that lead levels were a potent predictor of criminal behavior. Whether such studies will stand up after additional research on larger samples and to what extent lead accounts for crime trends on a macro level remain to be seen.[19]

• • •

We have discussed the two most prevalent sources of lead exposure in the modern era: house paint and gasoline. Next is drinking water, which has been a problem for much longer than the first two. In Roman times, lead lined the aqueducts that carried water into the cities and also was present in the network of lead pipes through which water was distributed within cities. In more modern times, lead piping was phased out and replaced by iron pipes. However, iron piping tended to rust and become blocked. Copper then came to the fore because it did not rust and was relatively light. However, copper pipes had to be connected together, and the best and most cost-effective manner of joining them required the use of lead solder.

In 1989, female reporters at the national newspaper USA Today had become concerned because a number of them had had miscarriages in 1987 and 1988. All of these women worked together on two floors of a twenty-two-story office building in Rosslyn, Virginia. In December 1988, the National Institute for Occupational Safety and Health (NIOSH) received a request for a health hazard evaluation from USA Today. The NIOSH found that most of the miscarriages had occurred among the eight reporters on one floor of the building who became pregnant between May and September 1988. The difference between this 100 percent miscarriage rate and the rate for the other parts of the building was statistically significant, making this a bona fide miscarriage cluster. The problem was identifying a cause. The newspaper's employees attributed the miscarriages to a variety of occupational factors, including exposure to chemicals used during a recent renovation of the workplace, the use of video display terminals (VDTs), and psychological stress.

The NIOSH did not confirm any of the theories advanced by employees but did detect the presence of lead in their drinking water. However, the level of lead in the water on the USA Today floors actually was lower than on the other floors of the building. Furthermore, lead was not detected in the blood of most of the women in the miscarriage cluster. The NIOSH concluded that the available studies linking miscarriages to lead were inconclusive, and, moreover, none of the USA Today reporters had evidence of lead exposure.[20] Although NIOSH ruled out lead, it failed to find a cause for this apparent cluster of miscarriages.[21] People in the building remained concerned about their lead exposure, especially the reporters who worked for USA Today.

Washington Occupational Health Associates, whom I worked for at the time, was contacted by the Gannett Corporation, who owned the newspaper. We found that the source of lead contamination in the water of the *USA Today* offices seemed to be near the tap, given that when the water was first turned on the amount of lead was highest and then decreased the longer the water ran. Most of the soldered joints are close to the tap, so it appeared that the problem was caused by lead used in the solder connecting the copper pipes. Solder containing lead had not been banned until the Safe Drinking Water Act Amendments of 1986, and this building had been constructed before that date.

The treatment of the lead problem in this building was a challenge. It would have been possible, but quite costly, to remove all of the water pipes and replace them with new feeds that did not contain the lead solder. We recommended another solution, a water treatment system inserted into the main water feed to coat the insides of the piping so that the lead would not go into solution. We also had the incoming water treated with alkali to make it less acidic, as lead is dissolved more readily in acidic water. The problem was solved: further testing showed barely detectable levels of lead in the tap water.

• • •

In the late 1980s it appeared that America's lead problem was well on its way to being solved thanks to all that we had learned and applied through the extensive government programs at the city, state, and federal levels. Not until the Flint, Michigan, water supply scandal hit the papers in 2015 were we jolted out of our complacency. The shocking revelations out of Flint made it abundantly clear that lead contamination remains a serious threat to public health in this country.

Flint's nightmare began when public officials switched the source of the municipal water supply from Lake Huron to the Flint River to save money. The more acidic nature of the Flint River water caused leaching of the lead from pipes at a higher rate than had occurred with the Lake Huron water. This was a risk that should have been obvious to anyone in the city or state government with a science background. Simultaneously and without any apparent forethought, the corrosion control system long in use in Flint was discontinued. This had the effect of exposing the lead in the pipes directly to the water. This double whammy boosted the lead in the drinking water to levels toxic for children.[22]

The Flint scandal called attention to ongoing lead contamination issues in other cities, most notably in Cleveland, the largest city in the neighboring state of Ohio.[23] Testing of water at Cleveland schools showed lead levels so high as to require the replacement of hundreds of drinking fountains, faucets, and other fixtures in sixty buildings.[24] Lead paint also remains pervasive in the city. While 7 percent of the children in Flint were found to have elevated levels of lead in their blood, the comparable figure for Cleveland is 14 percent. Given that Cleveland's population is four times Flint's, the total number of at-risk children is eight times greater.

The scale and geographic scope of America's lead threat is daunting. In the country as a whole, 24 million homes and apartments contain potentially hazardous levels of lead in soil, paint chips, or household dust, according to the estimates available from 2014, and four million of these households have children. In just three East Coast cities—Atlantic City, Philadelphia, and Allentown—five hundred children are believed to have enough lead in their blood to require a doctor's attention.[25]

Yet another form of lead contamination also has come to public attention recently. The lead that was used to make paint and pipes was produced throughout the country in smelters, many of which were located in densely populated areas. These facilities were closed down years ago but left behind acres of soil contaminated by lead. Although not as pervasive a problem as the lead in paint and pipes, there are hundreds, if not thousands, of former smelting sites in the United States.[26] One drew national attention recently when the mayor of East Chicago decided to raze a housing project and relocate its residents after learning that the soil there had been contaminated with lead from a nearby defunct smelter. Apparently, the EPA had narrowly concentrated on the cleanup of the industrial site and did not investigate the condition of the surrounding neighborhood. Officials from the city, state, and the U.S. Environmental Protection Agency resorted to finger pointing rather than accept responsibility for having failed to inform local residents about the risk to their children.[27]

The problem of tracing lead contamination to its source does not end with paint, soil, drinking water, and dust containing the fallout from leaded gasoline and smelters. Cans and other lead-containing vessels also exposed children to unhealthy levels of lead. Lead solder was used to make cans to store evaporated milk and fruit juices. In the early 1940s, this solder contained 63 percent lead and 37 percent tin, but during a tin shortage in World War II, the lead content rose to a supertoxic 98 percent. In the early 1970s, high levels of lead still were being found in products stored in cans and intended for consumption by children. As a result, the Food and Drug Administration took steps to decrease the amount of lead in the solder and also promoted the use of other types of containers for milk and fruit juices.[28]

The list of other potential exposures to lead in any given household is quite long. Many non-Western folk remedies used to treat diarrhea or other ailments contain substantial amounts of lead. In a study published in 2001, lead was the most common heavy metal contaminant/adulterant found in samples of Asian traditional remedies sold at health food stores and Asian groceries in Florida, New York, and New Jersey. Fully 60 percent of the remedies tested imparted a daily dose of lead in excess of 300 mg when taken according to labeling instructions.[29]

• • •

Studying lead poisoning in children teaches us three important lessons about toxicology. One is that there may be a wide variety of sources of exposure for a poison, and this makes it difficult to determine the cause of poisoning in any individual. It took many decades after lead poisoning was first identified in children to understand the complexity of the problem.

Another crucial lesson concerns how the small exposures, initially thought to be trivial in children, became important once neurological testing became more sophisticated. We know that the dose makes the poison, but it is equally important to be able to tell when the poisoning is occurring, and our information is always limited by the sensitivity of our tests. A related observation is that studies that correlated lead in large groups of children with the results of their neurological tests provided significantly more robust information than observations of test results in individual children.

Setting aside these purely scientific concerns, we must acknowledge that public health problems like lead poisoning tend not to be effectively addressed unless there is persistent pressure put on politicians and government officials by the public and by medical professionals, including toxicologists. Otherwise, apathy takes over, and the will to do something about a long-term problem like lead dissipates. Other more immediate priorities take precedence, funding gets shifted to other things, and the children carrying the burden of lead in their bodies are forgotten.

There is blame aplenty for government bureaucrats, elected officials, and even toxicologists for America's protracted failure to protect its children—especially those living in disadvantaged inner-city neighborhoods—from poisoning by lead. "We know how to fix it," says David Jacobs, who was director of the Office of Healthy Homes and Lead Hazard Control at HUD from 1995 to 2004. "The technology is there. It's just a matter of political will to properly appropriate the money."[30] Although Dr. Jacobs was referring to the lead-based-paint problem, his statement also applies to lead in drinking water, soil, and other sources. We do indeed have the technology to protect our children, but the fixes are expensive and require steady, persistent effort over many years.

8 | Rachel Carson

Silent Spring Is Now Noisy Summer

here was one other source of lead in the environment: a lead compound that caused widespread contamination of agricultural products. Invented for use as a pesticide, lead arsenate ($PbHAsO_4$) was first sprayed in 1892 against the gypsy moth in Massachusetts. A few years later, apple growers began using it to combat the codling moth, a destructive insect pest. Lead arsenate is a combination of two very toxic agents—lead and arsenic—but it was popular among farmers nevertheless because it was immediately effective, cheap, easy to use, and long lasting. Over the next sixty years the frequency and amount of lead arsenate applications increased, leading eventually to the development of pesticide resistance. There then came a downward spiral of decreased efficacy, requiring growers to keep increasing rates and application frequency.[1] Clearly, something new was needed.

Enter dichlorodiphenyltrichloroethane, or DDT, the first completely synthetic pesticide, which was invented by the Swiss chemist Paul Herman Muller in 1939. Muller began developing insecticides in 1935, realizing that the United States alone was using 10 million pounds of lead arsenate yearly. Muller had tested

349 chemicals by spraying them on flies in a cage. The 350th chemical that he tried produced a knockout punch: all the flies fell helplessly on their backs in minutes or hours. At first, Muller's peers at the J. R. Geigy Corporation in Basel, Switzerland, were unimpressed; lead arsenate produced almost instantaneous insect death as well. But DDT produced its lethal effect consistently, and the sprayed cages remained lethal environments for months or even years. It didn't target flies only: this new chemical proved deadly to a wide variety of insects, including mosquitoes, aphids, beetles, and moths. Eventually Geigy was convinced and took out a patent on the use of DDT as an insecticide in 1940,[2] and Muller was awarded the Nobel Prize in physiology and medicine in 1948 for his discovery of DDT.[3]

With the outbreak of World War II, DDT quickly became indispensable. In 1943, four hundred pounds of DDT arrived by freighter in New York City and were sent for testing at the Department of Agriculture and the Kettering Laboratory at the University of Cincinnati, College of Medicine. They determined it would be safe and effective in battlefield situations to fight typhus, malaria, yellow fever, and encephalitis. Field testing in U.S.-occupied Naples in 1943 showed DDT to be far superior to the existing treatments for lice. Ninety percent of the population was infested, resulting in a typhus epidemic whose death rate approached 25 percent.

DDT was hailed as a miracle substance on a par with penicillin. Millions of people were sprayed or dusted with DDT during the war and after with no apparent harm. Soon DDT was being sprayed everywhere—in homes, barns, forests, and croplands—replacing the highly toxic insecticide lead arsenate.[4]

In the mid-1950s, the World Health Organization (WHO) had made the eradication of malaria a priority, and DDT was its major weapon.[5] Use of the chemical grew even though the WHO had found that resistance was developing to the insecticide.[6] Malaria killed hundreds of thousands of people every year and sickened millions. Legislation authored by the senators John F. Kennedy and Hubert Humphrey gave $100 million to the World Health Organization malaria project, which was using DDT. By 1960, malaria had been eliminated in eleven countries. In India, the number of cases was cut from 75 million people to below one hundred thousand.[7] In the meantime, back home the use of DDT grew at a rapid rate. In 1950, only about one-tenth of the DDT produced in the United States was shipped overseas; the rest was consumed domestically. Americans used 80 million pounds of DDT in 1959 alone.[8]

When lead arsenate and DDT were introduced, there was little regulation of pesticides. The Federal Insecticide Act (FIA) of 1910 merely prohibited false labeling of products. In 1947 Congress passed the Federal Insecticide, Fungicide, and Rodenticide Act (FIFRA), but it was concerned with the efficacy of pesticides; it did not regulate pesticide use. The 1954 Miller Pesticide Amendment to the Food, Drug, and Cosmetics Act would set limits on pesticide residues on raw agricultural commodities, but it would not be until 1972, with the passage of the Federal Environmental Pesticide Control Act, that the use of pesticides would be restricted to protect human health and the environment.[9]

Meanwhile it was becoming clear that DDT and the other, more powerful synthetic organochlorine pesticides, such as lindane, chlordane, and heptachlor, and organophosphate pesticides, such as parathion, were toxic to wildlife and therefore potentially harmful to humans. A zoologist named George J. Wallace noticed a die-off of robins at Michigan State College in the spring of 1955. Each spring, usually following the spraying of the elms for Dutch elm disease, many birds would exhibit symptoms of poisoning. By 1958, robins had been eliminated from the main campus and portions of East Lansing. In all, forty species of birds were found to be affected. Richard F. Barnard detected DDT residues in the tissues of the dead birds at Michigan State University, which strengthened the link between the deaths and the application of the pesticide DDT to control elm bark beetles.[10] Like lead, organochlorine chemicals such as DDT are not easily eliminated from the body, which makes it a relatively straightforward task for scientists to associate exposures to blood or tissue levels to the effects of these chemicals on the body.

In the fall of 1957, fourteen residents on Long Island in New York State filed a lawsuit against the federal government to prevent the spraying of DDT to control gypsy moths. It was reported that a spraying program for gypsy months on Long Island had killed 83 percent of the bird population in two weeks.[11] The suit caught the attention of Rachel Carson, a wildlife biologist formerly with the U.S. Fish and Wildlife Service who had been concerned about pesticides for some time and routinely clipped newspaper articles on pesticide issues. When the trial started in early 1958, Carson's interest became known to one of the plaintiffs, Marjorie Spock, who then contacted her. That summer, Spock forwarded to Carson a large file of materials from the lawsuit, which claimed that the spraying contaminated milk in dairy cows and posed a threat to people's health. An expert witness in the trial on behalf of the plaintiffs was Dr. Malcolm Hargraves, from the Mayo Clinic. The state of New York said that it was canceling the spraying program for 1958, but the court ruled that the spraying program was legal and successful. Carson's interest in the lawsuit and some of the documents that she had obtained grew into a writing project that would eventually become the landmark book *Silent Spring*.[12]

There were other influences on Carson's work. On October 7, 1958, the Harvard evolutionary biologist Edward O. Wilson wrote Carson a letter after he heard about her studies and planned book. Elton's *Ecology of Invasions* had just been published, and Wilson suggested that Carson consider this work. Carson found this book enormously stimulating: "It cuts through all the foggy discussion of insect pests and their control like a keen north wind."[13]

In late 1959, there was a major human cancer scare caused by chemical food contamination. Cranberries were found to be contaminated with the herbicide amitrole, which had been shown to cause thyroid tumors in rats. Most of the thyroid tumors reported were considered to be benign, but five of the tumors were interpreted by some of the pathologists as cancerous.[14] It had been banned earlier in June 1959 by the Delaney Amendment, which had been passed the previous year. Only cranberries from Washington and Oregon were thought to be involved, but consumers refused to buy any from Wisconsin, Massachusetts, or New Jersey, either. Later, the Wisconsin berries were found to be contaminated as well. Thanksgiving and Christmas dinners lacked this ingredient that year, and furious growers were calling for the resignation of Arthur Flemming, the secretary of the Department of Health, Education, and Welfare.[15] Amitrole was banned, even though no one knew whether it had any health effects in humans. In a stunt organized by a local radio station in Massachusetts, a crowd of thousands drank a thousand gallons of cranberry juice. Senator John F. Kennedy was in Wisconsin and drank a cranberry juice toast. Not to be outdone, Vice President Richard Nixon ate four helpings of cranberries, even though Secretary Flemming advised against it.[16] The cranberries contaminated with amitrole would play a prominent role in the chapter on human cancer in Carson's book.

Silent Spring, published in 1962, was a wake-up call for scientists to evaluate critically whether cancer was caused by industrial pollution. Carson was a respected scientist and an excellent writer, yet there was something evangelical about this book. Carson had considered calling it *Man Against the Earth*, but when it became clear that the mass death of birds was a major theme, she chose the title *Silent Spring*. It was one of the first books written for a general audience that introduced toxicological concepts. It was selected for the Book-of-the-Month Club in October 1962 and would go on to become a major bestseller.[17]

In *Silent Spring*, Carson recounted numerous instances when the spraying of a pesticide led to the death of hundreds or thousands of animals. She documented the observed effects of the organochlorine insecticides chlordane, DDT, DDD, heptachlor, endrin, and dieldrin and organophosphates like parathion.[18] She reported sickness and death in birds, fish, amphibians, reptiles, and mammals. Since insecticides are designed to kill insects, it was to be expected that the loss of this food source would cause mortality for birds and fish. But there was also evidence of a direct, toxic effect on wildlife and effects progressing up the food chain.

One example Carson described was a campaign to eliminate Japanese beetles in 1954 with dieldrin in Sheldon, Illinois. After the aerial spraying, robins and other birds were virtually wiped out. Ground squirrels, muskrats, rabbits, and some sheep also died. The campaign was continued the following year using aldrin, with similar results on the wildlife. Another example was the spraying of the Miramichi River in New Brunswick, Canada, in 1954. The target was the budworm, which attacks several kinds of evergreens. DDT was used on millions of acres of forests, and the result was the death of salmon and brook trout. The Fisheries Research Board of Canada had been studying the river since 1950; the kills of the salmon were well documented. Fortunately for the salmon, in this case a severe storm washed away most of the contamination, and some of the fish recovered.

• • •

The close temporal relationship of acute toxic effects following the spraying of organochlorine insecticides—in some cases within days—provided good evidence of a causal relationship. This relationship was strengthened by the sheer number of cases, with large populations of affected animals, different geographical locations of the episodes, and various environmental situations. These seemed to rule out the possibility of coincidence.

Carson was a wildlife biologist, so she was on firmer ground when writing about animals. She took a chance in *Silent Spring* in extending her argument to human health. She did not take this chance lightly. She knew that she would have to be more speculative in discussing human cases, and she considered avoiding the subject altogether. But Carson knew that readers would inevitably make inferences from her observations in wildlife to the possible implications for humans. Plus, she was suffering from breast cancer during the time she was writing *Silent Spring*.

Carson struggled with how to handle the information on cancer and how much to emphasize it. In a letter to Paul Brooks, who was editor-in-chief of Houghton Mifflin's general book department and who would publish *Silent Spring*, she wrote: "Until recently, I saw this as part of a general chapter on the physical effect on man. Now it looms so terrifically important that I want to devote a whole chapter to it—and that perhaps will be the most important chapter of the book."[19] Having decided to take the plunge into human health, she proceeded in a careful, stepwise manner. First, in her chapter "Beyond the Dreams of the Borgias," Carson chronicles a number of acute poisoning incidents in humans where the temporal relationship is strong. Next came the chapter on cancer, called "One in Four," in which she poses the following question: "The problem that concerns us here is whether any of the chemicals we are using in our attempts to control nature play a direct or indirect role as causes of cancer."[20]

Carson begins her cancer chapter by describing the use of arsenic compounds as weed killers and insecticides. She cites Wilhelm Hueper's 1942 publication *Occupational Tumors and Allied Disease*, which describes the studies of miners of Reichenstein, Silesia, who were exposed to arsenic ores of gold and silver and suffered various ailments, including cancer.[21] She also describes skin cancers from naturally occurring high arsenic levels in drinking water in Cordoba Province in Argentina. She does not describe any cancer among users of arsenic-containing pesticides but states that "it would not be difficult to create conditions similar to those in Reichenstein and Cordoba by long continued use of arsenical insecticides." This is not much of a leap, of course. After all, Paul Herman Muller had replaced the extremely toxic and widely used arsenical insecticides with DDT.

Carson next turns her attention to DDT, which seems to be her main concern. In 1952, Hueper, from the National Cancer Institute, testified to the congressional Miller Committee that he could not call DDT carcinogenic for humans and that it would take ten to fifteen years before such a case could be made.[22] Carson's evidence for the carcinogenic effects of DDT was largely circumstantial and based on the unpublished clinical observations and opinions of Hargraves, the plaintiffs' expert witness in the Long Island trial. Carson reported that Hargraves routinely saw patients with leukemia and lymphoma in his practice at Mayo whose illnesses he believed were caused by pesticide exposures. However, he conceded that he had not verified his beliefs through any rigorous research.[23] According to a recent biography of Rachel Carson, she had been warned by the cancer expert who became her oncologist at the Cleveland clinic, Barney Crile, that the link between pesticides and human cancer was scary but speculative. This was especially the case because making that link relied on a primitive understanding of cell biology and on anecdotal observations, such as those of Hargraves, whose testimony about the connection between pesticide use and leukemia during the Long Island case had so impressed Carson.[24]

Only one case of malignancy associated with exposure to DDT was specifically described to support Carson's claim of a link to cancer in her book. This was one of Hargraves's patients, whom Carson described as a housewife afraid of spiders and who sprayed with DDT. She was found to have acute leukemia, apparently within a couple of months. Such a short latency, however, is unknown for cancer. Carson acknowledges that it takes several years to develop leukemia following a chemical exposure. So why did she ascribe any credibility to this report?[25] She refers to other cases of exposures to pesticides from the literature and Hargraves's patients, but she offers no specific descriptions of them in her book. In a presentation by Hargraves to the National Wildlife Federation, all other cases of leukemia described by Hargraves involved petroleum distillates or other pesticides but not DDT.[26]

Carson knew that she had to dig deeper, and she tried to bolster these findings with arguments from biological plausibility. She wrote in her letter to her editor, Paul Brooks:

> To tell the truth in the beginning I felt the link between pesticides and cancer was tenuous and at best circumstantial; now I feel it is very strong indeed. This is partly because I feel I shall be able to suggest the actual mechanism by which these things transform a normal cell into a cancer cell. This has taken very deep digging into the realms of physiology and biochemistry and genetics, to say nothing of chemistry. But now I feel that a lot of isolated pieces of the jig-saw puzzle have suddenly fallen into place. It has not, to my knowledge, been brought together by anyone else, and I think it will make my case very strong indeed.[27]

She chose a theory by the German biochemist Otto Warburg, which she believed provided the mechanism whereby pesticides caused cancer. According to Carson's interpretation of Warburg's theory for pesticide toxicity, pesticide exposure damaged the cell's ability to produce energy in the form of ATP, which is the body's main source of oxygen-derived cellular energy. Carson writes that this causes a "grueling struggle to restore lost energy." The cell compensates for this lost energy by increasing fermentation, and "at this point, cancer cells may be said to have been created from normal body cells."[28]

Otto Warburg was doing Nobel Prize–winning work on cell metabolism, but these were still the early days of biochemistry. He had discovered a correlation in cancer cells between decreased mitochondrial respiration and increased fermentation, but correlation between two events does not demonstrate cause and effect. It turns out that the metabolic changes Warburg described were an effect of rather than the cause of cancer. Cancer cells, which have already been produced by alterations in the genome, thrive under fermentation energy production rather than normal mitochondrial respiration.[29] This is known as the Warburg effect and is the basis of today's PET scans used in cancer detection. However, there is little or no support for the notion that the lack of mitochondrial respiration is a cause of cancer.

• • •

The response of the agricultural chemical industry was swift, coming hard upon the publication of the first half of *Silent Spring* in three issues of the *New Yorker*. A *New York Times* article titled "*Silent Spring* Is Now Noisy Summer" announced that the "drowsy midsummer has suddenly been enlivened by the greatest uproar in the pesticides industry since the cranberry scare of 1959."[30] The pesticide manufacturers were crying foul: "crass commercialism or idealistic flag waving"; "We are aghast"; "Our members are raising hell." There was little dispute of the facts in her book, but pesticide manufacturers strongly objected to implications that industry and the government were being careless and destroying the environment. This was weeks before the complete book had been released.

The federal government acted soon afterward. In 1963, the year following the publication of *Silent Spring*, the Food and Drug Administration gave special attention to the pesticides aldrin and dieldrin, which had been found to cause tumors in mice.[31] Aldrin had been detected in concentrations above the Food and Drug Administration's tolerance levels on potatoes in the Pacific Northwest. The government officials involved were quoted as saying, "If Rachel Carson overstated the case, she erred on the side of the angels." Of course, the implication was that the other side, industry, was not on the side of the angels. In May 1963, President Kennedy's Scientific Advisory Committee asserted that pesticides were needed to maintain the food supply but warned against their indiscriminate use. According to Carson, this was precisely the point of her book.[32]

It's important to note that nowhere in *Silent Spring* did Carson argue against the use of DDT to fight disease.[33] Both her critics and supporters seemed to have ignored the care with which she presented her opinions on this subject. After the second installment of her book came out in the *New Yorker*, she was attacked incorrectly for being against the use of pesticides to control disease.[34] However, for the protection of wildlife, she probably would have supported the actions by Secretary of the Interior Stewart Udall, who instituted tough new rules restricting pesticide use on more than 550 million acres of federal lands in 1964.[35]

Rachel Carson died in 1964 of cancer, but the furor caused by her book only escalated. Eventually in 1970, President Richard Nixon created the Environmental Protection Agency (EPA), and one of the first actions it took was to cancel the pesticide registration (but not the manufacture for export) of DDT, aldrin, dieldrin, chlordane, heptachlor, and endrin.[36]

Although Carson's historical importance is assured, many of her scientific assertions about human health effects have not been borne out. In her discussion of the cranberry scare over the herbicide amitrole, Carson cited findings in which 100 parts per million of amitrole produced thyroid tumors in the rats.[37] The EPA banned all food crop uses for amitrole in 1971. However, even in their latest reviews, amitrole still has not been found to cause cancer in humans by the EPA or by the International Agency for Research on Cancer (IARC).[38]

Carson had implicated DDT in the development of aplastic anemia and bone marrow wasting, but according to the most recent review of studies in humans by the Agency for Toxic Substances and Disease Registry in 2002, bone marrow, which is the target for the production of aplastic anemia, does not appear to be a sensitive target for DDT.[39] Numerous epidemiological investigations have studied people exposed to organochlorines, yet no conclusion regarding human cancer has been established. The IARC, the EPA, and the U.S. National Toxicology Program have all refrained from classifying DDT or any of the other organochlorines, such as aldrin, dieldrin, chlordane, heptachlor, and endrin, as human carcinogens.

And the debate continues as to whether the benefits of using DDT outweigh the risks. In recent years there have been resurgences in the use of this powerful pesticide for malaria control, with a number of success stories.[40] In Suriname, DDT applications reduced the number of registered malaria cases from 14,403 in 2003 to 1,371 in 2009. Amir Attaran and Rajendra Maharaj from the Center for International Development, Harvard

University, and the South Africa Department of Health argue that DDT has saved millions of human lives, while the claims of adverse health effects have been based only on animal studies.[41] An article by Lorenzo Tomatis, the former head of the IARC monograph program, and colleagues argues that, although there were concerns about the health impacts of DDT, its usefulness for mosquito control in poor countries would preclude a ban of this pesticide.[42] On the other side, Richard Liroff of the World Wildlife Fund cites continuing impacts on wildlife and suggestive studies in experimental animals.[43]

In an ironic twist, it appears that not only humans but wildlife as well may benefit directly from malaria control. It had been thought that malaria does not affect wildlife, because many animals have become resistant to the parasite, but recent studies have questioned this presumption. A three-decade study of great reed warblers found that malaria-infected birds had shortened lifespans, which cuts their reproductive lifespan to approximately half of that in uninfected birds.[44]

Decades after the publication of *Silent Spring*, controversy about the use of DDT and other pesticides continues, countless ecological and medical studies are performed, and regulations are written and revised. All of this can be regarded as Rachel Carson's legacy. Certainly, thanks to her observations, DDT and many other pesticides will never be indiscriminately used again.

9 | The Study of Cancer

Paracelsus studied only humans. He threw out the Gallenistic concepts of medical treatment and based his remedies solely on his observation of his patients. Likewise, when he studied the diseases of miners, he focused solely on the symptoms and signs that he saw. His description of *mala metallorum* was probably lung cancer caused by mining arsenic or radioactive dust. But he didn't understand the exposures of those miners well enough to identify the causative agents.

The next discovery of human cancer caused by chemicals would come from a study of a disease affecting an unlikely place, the male scrotum. In 1775, a year before the American Revolution began, Sir Percival Pott, at St. Bartholomew's Hospital in London, reported the occurrence of scrotal cancer among chimney sweeps in London. At that time it was common practice throughout Europe to employ children as chimney sweeps. Pott wrote that "in their early infancy, they are most frequently treated with great brutality, and almost starved with cold and hunger; they are thrust up narrow, and sometimes hot chimneys, where they are bruised, burned, and almost suffocated; and when they get to puberty, become peculiarly liable to a most noisome, painful, and fatal disease." He continues, "but it is nevertheless a disease to which they are peculiarly liable; and so are chimney-sweepers to the cancer of the scrotum and testicles." He hypothesized that soot adhering to the skin folds in the scrota of chimney sweeps was causing scrotal cancer.[1]

More than a century later, Henry Butlin, also from St. Bartholomew's Hospital in London, wrote "Three Lectures on Cancer of the Scrotum in Chimney-Sweeps and Others." In the first lecture, published in 1892, he noted that cancer of the scrotum rarely occurred outside of England. In Berlin during the eight years from 1878 to 1885, there was not registered a single case. No cases were found in Paris in the year 1861 and only one case in the City Hospital in Boston from 1881 to 1887. However, there were still reports of cancer of the scrotum in London during this time period.[2] Published lectures reported that personal hygiene, including washing, was not practiced by chimney sweeps in England. In contrast, the relative absence of scrotal cancer among chimney sweeps in other parts of Europe could be explained by better hygiene, including bathing and the use of clean clothes. Butlin concluded that the cause of the cancers was the tar in the soot that clung to the scrotum.[3] However, the exact nature of how the chemicals in tar caused cancer would have to wait until the chemicals were identified and the process of cancer development was further elucidated.

• • •

The physician Marie François Xavier Bichat (1771–1802), at the Hotel-Dieu, considered the oldest hospital in Paris, developed the idea that organs were composed of basic types of tissues, which were variously distributed in the organs. Cancers are often described in terms of the organ of origin, but it is a particular tissue within the organ where cancers arise. Bichat studied the development of tissues not with a microscope but by observing the effects of chemical tests on organs. For example, the outer layer of the skin, the epidermis, contains tissue called the epithelium, which typically has both an outer surface and a fibrous layer underneath called the basement membrane. Under the basement membrane is the dermis, which is the skin's connective tissue. The dermis has the blood vessels, lymphatic system, collagen, and other fibers that keep the skin in one layer and connect it with the deeper muscle and nervous tissues. The stomach and intestines of the digestive system are more complex in their tissue compositions. The mucosal layer is in contact with the decomposing food and contains layers of epithelial, connective, and muscle tissue. Below is a submucosal layer of connective tissue that contains the blood vessels, and below that is the muscularis externae, or major outer muscular layer, which propels the food through the digestive tract. Finally, there is the serosal layer of connective tissue, which forms the outer layer, in contact with the peritoneal space.

Another organ Bichat focused on heavily was the heart. He found that there were three different tissue types and that they were subject to three different types of inflammatory disease: pericarditis, myocarditis, and endocarditis, involving the outer serosal connective tissue, the muscle, and the inner connective tissue containing the endothelium, respectively. He also subjected the organs to chemical tests—soaking, drying, and decomposition, among other means—to differentiate the different tissues.[4] Bichat succeeded in moving the focus of disease from the organ as a whole to its various tissues. This led the way to understanding that certain diseases could affect different organs that had the same types of tissues.[5]

The prevailing view at the beginning of the nineteenth century was that tumors consisted of clotted and degenerated lymph and that the growth was merely a local manifestation of a general disease analogous in this respect to the pustule in smallpox. This would change with the invention of the achromatic microscope in 1824 by Johannes Müller. Achromatic lenses are corrected to bring two wavelengths (typically red and blue) into focus in the same plane. Previously, there were distortions caused by the different wavelengths of light focusing in different places. As a pathologist studying human cancer, by 1830 Müller had become an experienced histologist; in 1838 he published the first extensive microscopic study of diseased tissue. This work, *On the Nature and Structural Characteristics of Cancer*, forms the foundation for the modern conception of the nature of cancerous growths.[6] His monograph on cancer provided a systematic analysis of the microscopic features of benign and malignant human neoplasms.

Müller utilized Bichat's understanding of tissues to demonstrate that cancer consisted of an abnormal growth of cells. Müller demonstrated that cancer is cellular and that the cellular form of a particular cancer resembles that of the tissues from which the growth originates. He clearly differentiated a malignant cancer from a benign tumor; the former has the ability to regenerate itself after removal. Microscopically he described cancer as an abnormal growth of cells that retain the characteristics of the cells from which they arose. As the founder of microscopic pathology, Müller inspired an entire generation of German scientists.[7]

• • •

Eventually, research on animals would become important to our understanding of how chemicals can cause chronic human diseases, including cancer, because experiments could be done within definite timeframes and involving regulated doses. But the interpretation would be much less straightforward than isolating infectious agents or describing the deadly effects of venoms. Before experimental research on the causes of cancer could proceed, scientists needed to figure out what constituted cancer in its various manifestations.

Plant biologists had previously discovered the binary nature of cell division, whereby one plant cell propagates another. Pupils of Müller went on to define the cellular nature of animal life. At first, two of these scientists, Matthias Schleiden and Theodor Schwann, incorrectly developed the theory that new cells were derived from the "cytoblast," which is the cell nucleus, and that the nucleus was derived from extracellular material. However, Robert Remak, who was also a pupil of Müller, rejected this theory, which appeared to him no different from the theory of spontaneous generation. His microscopic observations beginning in 1841 on the development of the chick embryo led him to propose that binary cell division in animals was the source of new cells. His various publications from 1845 to 1854 were initially rejected by Rudolf Virchow, another pupil of Müller. But finally in 1858 Virchow was won over to Remak's ideas and published "Cellularpathologie," where he popularized the idea of binary cell division. He stated "Omnis cellula e cellula," which means "all cells come from cells." This phrase became the catchphrase for the primacy of the cell in life.[8]

The first study that pointed to chromosomal changes in cancer cells was performed in 1890 by the German pathologist David von Hansemann, who had been an assistant of Virchow, by studying human cancer biopsies using staining techniques that allowed the study of cell nuclei. Hematoxylin, which stains the chromatin in the nuclei of cells, and eosin, which stains the cytoplasm, were developed in the 1860s and widely used by 1880. Hematoxylin was produced from extracts of logwood from southern Mexico and Central America; eosin was a synthetic aniline dye. Von Hansemann noted the frequency of abnormal-looking nuclei in cancer cells undergoing replication, which is known as anaplasia. Such cells are called anaplastic. He proposed the theory that normal cells became cancer cells because of a primary alteration in the hereditary material.[9]

There were seven components of Hansemann's theory, and three of these in particular have withstood the test of time. First, Hansemann said that cancer cells lose the specialized functions of normal cells. Second, cancer cells have a capacity for independent existence in tissues. And finally, there was a loss of differentiation, which he described as part of anaplasia. He also described cancer cells as becoming more and more bizarre in their cellular and nuclear morphology as they became metastatic in organs beyond their site of origin.[10]

The next important contribution came from an experimental zoologist studying sea urchins. In his seminal book, *Concerning the Origin of Malignant Tumours*, published in 1914, Theodor Boveri described several important aspects of the nature and origins of malignant tumors. In 1902 he had expanded on the results of his experiments on the development of doubly fertilized sea urchin eggs to speculate that malignant tumors might be the consequence of a certain abnormal chromosome constitution.[11]

Acknowledging the important discoveries of Hansemann as closely related to his own, Boveri described the anaplastic cell's altered state as reacting differently to its environment. According to Boveri, malignant cells often resembled cells of tissues different from their site of origin. He proposed that this altered property resulted in a tendency to multiply without the usual restraint of nonmalignant cells. Boveri concluded from his observations that the defect of malignant cells was located in the nucleus and not the cytoplasm. He made the strikingly prescient hypothesis that tumors arise from a defect in a single cell that is propagated and even maintained in metastasis.[12]

• • •

Based on the studies by von Hansemann and Boveri, cancer was thought to involve chromosomes. In the late 1940s, it was suspected that some chemicals must be capable of altering the chromosome's genetic code; the question was how. Charlotte Auerbach and J. M. Robson at the University of Edinburgh were the first scientists to demonstrate the foundation of the mutations caused by a chemical—in this first case, the infamous mustard gas used in World War I. They already knew from experiments with flies that radiation caused mutations and sterility. The mustard gas that they tested in 1941 had similar effects. To explore further the mechanism of the chromosomal mutations, they designed a special study to see whether the altered DNA could be seen under a microscope.[13]

To study chromosome changes, investigators borrowed a technique that had been developed to study plants and insects. When we think of genetic material, we usually picture chromosomes as cigar-shaped structures, which is how they look when they have been compacted for cell division during mitosis. Otherwise, chromosomes are long strings that are not visible under a microscope. When scientists look for genetic damage, they look at the chromosomes during one part of the cell replication process called metaphase, when the duplicated chromosomes are all lined up before segregation into the two daughter cells. This metaphase analysis of chromosomes is the discipline called cytogenetics.[14]

It was not until 1956 that the correct number of forty-six human chromosomes in twenty-three pairs in a cell was identified.[15] During cell division in somatic cells, each needs to be replicated in order to produce two daughter cells. For most of the cell replication cycle, each chromosome contains long strands of DNA, each containing about a thousand genes, which have to be packaged into the cigar-shaped forms so that they are able to segregate into the two daughter cells during mitosis.

Usually when we think of inheritance, we think of genetic characteristics, also known as phenotypes, passed from parents to their offspring. An acquired mutation can be passed from generation to generation, and this is known as a "germline" mutation. An example of this is the altered hemoglobin in sickle cell anemia, which is passed from one generation to the next. More recently discovered examples are the mutated *BRAC1* and *BRAC2* genes, which confer a much greater risk of developing breast cancer and are passed from one generation to the next.

As later summarized in 1959 by Macfarlane Burnet, cancer caused changes that were inherited in "somatic" cells, which are any cells in animals that are not involved in the transmission of genes from one generation to the next by sexual reproduction. For cancer to develop, it would later be discovered that if a critical gene gets mutated in a somatic cell in one's body, this cell has the potential to become a cancer cell. This process eventually involves several additional mutations in the somatic cell line.[16]

The discovery of the structure of chromosomes and DNA provided the basis for understanding the molecular basis for the inheritance of somatic mutations during cell replication. It also led to understanding the nature of mutations and their inheritance in cancer cells. After studying the helical structure developed by Linus Pauling for proteins, Francis Crick and James Watson proposed a double-helical structure for DNA in 1953 based on the X-ray crystallography of Rosalind Franklin.

The backbone of this structure was two strands of alternating ribose sugars and phosphate groups with ringed purine and pyrimidine bases attached inside each of the strands, which formed weak hydrogen bonds holding the strands together. Although they first thought that like bases attracted like bases, this was discarded in favor of purines forming the hydrogen bonds with pyrimidines, so that guanine (G) was always opposite a cytosine (C) and adenine (A) opposite a thymine (T). This resulted in series of G-C and A-T base pairs forming the genetic code on complementary strands of DNA. When the strands separated, DNA replication of the complementary strands would occur, forming two new identical double-helical structures.[17]

Crick and Watson understood the implications of their model for the replication of DNA but did not appreciate the specific role of DNA in the cell. "Our model for DNA suggests a simple mechanism for the first process [replication], but at the moment we cannot see how it carries out the second one. We believe, however, that its specificity is expressed by the precise sequence of the pairs of bases."[18]

Francis Crick popularized the term "codon" in 1959, which had been coined by the biochemist Sydney Brenner to describe the fundamental units engaged in protein synthesis, even though the units had yet to be fully determined. Marshall Niremberg, at the NIH, along with J. Heinrich Matthaei, a young postdoctoral researcher from Germany, initiated a series of experiments using synthetic RNA. These two researchers were able to show how RNA transmits the "messages" that are encoded in DNA combining amino acids to make proteins. These experiments became the foundation of Nirenberg's groundbreaking work on the genetic code. Once they discovered the RNA "codeword" for the amino acid phenylalanine in 1961, these scientists set out to discover the unique codewords for all twenty major amino acids that combine to form proteins.[19] This information regarding the structure and process of DNA replication were fundamental to the understanding of cancer formation and propagation.

X rays and some chemicals such as mustard gas can cause extensive types of cytogenetic damage to DNA, for example, rupturing the backbone of the DNA, which results in the deletion of parts of chromosomes. Other effects involve the exchange—translocation—of genetic material between chromosomes. These types of chemicals are termed clastogens, and the damage can often be viewed under a microscope by cytogenetic analysis. In the analysis by Auerbach and her associates, translocations and chromosome breaks were the first types of damage caused by mustard gas detected in fruit fly chromosomes. These were the first confirmations that a chemical used during World War I (and still in existence during World War II) was capable of affecting genetic material, but their findings were not published until 1946, given wartime restrictions on describing studies involving chemical warfare agents.[20]

The most well-known translocation is the cause of chronic myelogenous leukemia, in which a part of chromosome 22 is exchanged for a smaller part of chromosome 9. Since chromosome 22 is already small, the resulting new chromosome 22 is even smaller. This was detected in 1960 by Peter Nowell at the University of Pennsylvania Medical School, and the new, abnormally small chromosome was called the "Philadelphia chromosome." This rearrangement puts two genes together that are not usually together in the sequence, and the resulting fused gene codes for a protein that causes the cell to replicate uncontrollably. As a result, there is continuous production of white blood cells in the bone marrow, resulting in leukemia. Even though chemicals can cause translocations in some circumstances, there are to this date no known chemical or other causes that produce the Philadelphia chromosome or chronic myelogenous leukemia.

10 | How Are Carcinogens Made?

B ased on this initial understanding of cancer in humans, experimental carcinogenesis used animals to explore further the development of the skin cancers originally described by Percival Pott. However, the next step would not involve chimney soot but rather a commercial product derived from coal. In 1845, the young German chemist August Wilhelm Hoffman came to London at the personal invitation of Prince Albert, the husband of Queen Victoria, to found the Royal College of Chemistry, with the aim of enhancing Britain's proficiency in practical chemistry. His main focus was extracting, analyzing, and testing compounds from natural sources, in particular coal tar, which was a coal-derived waste byproduct of coke production and coal gas lighting. Coke was necessary for iron production, and coal gas was the first major substitute for tallow candles and whale oil lamps. Coal tar had many uses, such as making pitch for waterproofing ships and roofs and creosote for preserving wooden railroad ties.[1]

The occurrence of skin tumors in coal tar workers had become well known thanks to the investigations of German doctors such as Richard von Volkmann (1830–1889). Skin cancer was also reported by Butlin among workers producing coal tar pitch, which was the source of gas, oils, and various chemicals that we now associate with petroleum but at that time were derived from coal.[2]

Scientists turned to experimental animals to understand the specific causes of coal-tar-induced cancer. Although no chemically induced carcinoma had yet been experimentally produced, Katsusaburo Yamagiwa and Koichi Ichikawa, at the Pathological Institute of the Medical College of the Imperial University, Tokyo, Japan, believed that cancer could be induced in experimental animals if the correct conditions could be found. They repeatedly painted crude coal tar on the ears of rabbits and were able to identify four distinct periods in the development of skin cancer that mimicked descriptions of cancer in humans. The first period was characterized by atypical growth of the epithelium, and the irritated ear soon became swollen. Under the microscope they could see mitotic figures, which were evidence of increased cell division, leading to an abnormal increase in the number of cells, or "hyperplasia." The second period was marked by the appearance of new growths, which they identified as benign tumors. In the third period, microscopic examination showed a high degree of proliferation of atypical cells with bizarre shapes or abnormally large and dark nuclei, indicative of carcinoma in its earliest stage. This developed further with the carcinomatous cells invading the surrounding subcutaneous tissue both downward and laterally, growing into the veins and the lymphatic channels, penetrating the ear cartilage, and forming ulcerated new growths. In the fourth period there was swelling of the regional lymph nodes, and these nodes were found by microscopic examination to contain metastatic deposits from the primary tumor. Other researchers followed suit in the laboratory with a number of coal tar distillates, including creosote.[3]

British researchers, primarily Ernest L. Kennaway (1881–1958) and his group at the Cancer Hospital Research Institute (CHRI) in London, made the most determined attempt to investigate tars for their various carcinogenic constituents. Exploiting recently established spectroscopic methods developed at the CHRI, James W. Cook and Kennaway's team in 1933 discovered benzo[a]pyrene, which is one of the polycyclic aromatic hydrocarbons (PAHs)—a large group of complex organic chemicals formed by incomplete combustion. This PAH was identified to be a principal carcinogenic component in coal tar. Other PAHs were purified from soot and coal tars, and their relative carcinogenic potencies were compared systematically through skin tests on mice.[4]

In a series of animal studies carried out at the University of Oxford in the 1940s, Isaac Berenblum and Philippe Shubik used benzo[a]pyrene and other PAHs in experimental animals to study the biologic mechanisms by which chemicals cause cancer. They showed that PAHs could be given in a single dose to irreversibly "initiate" normal cells into potential tumor cells. Other chemicals could be used to promote the development of these cells into tumors, but they needed to be dosed repeatedly for months. Specifically, a single application of benzo[a]pyrene, followed by repeated applications of croton oil, was sufficient to lead to tumor production. Croton oil is produced from the seeds of a tree and is very irritating to the skin.[5]

These studies presaged a major breakthrough in the understanding of how chemicals can cause cancer and predicted that there were different mechanisms by which different chemicals can cause cancer. The initiator chemical would eventually be found to cause genetic changes in cells by reacting with its DNA and causing a mutation, a genetic change that is passed on to the daughter cells. The promoter chemical, in turn, would be found to cause changes in the regulation of genes, thus altering the expression of a gene or its capacity to produce proteins.

• • •

In the early half of the twentieth century, it was not clear that chemicals caused cancer by inducing mutations, because some chemicals were mutagens but not carcinogens, but some carcinogens were not found to be mutagens.[6] As for mustard gas, which was the first chemical shown to cause mutations, there was not convincing evidence of its ability to cause tumors in experimental animals. Even in 1962, David Clayson had reported as the existing state of knowledge that many chemicals that caused cancer were not mutagenic and that a mutagenic chemical such as mustard gas was not carcinogenic.[7]

In the early 1950s at the National Cancer Institute, John and Elizabeth Weisburger, whom we met in chapter 6, were exploring the notion that certain chemicals known to be carcinogenic—including some drugs—are not reactive by themselves but are changed in the body into other entities that cause the cancers. Most chemicals that cause cancer are not inherently chemically reactive; otherwise, they would react with other chemicals in the environment and become inactive. There are some exceptions to this rule: one of them is formaldehyde, which is reactive and reacts directly on molecular components in the respiratory tract. But formaldehyde is the exception, and generally the quest became to figure out how seemingly inert organic chemicals like aromatic amines and polyaromatic hydrocarbons such as benzo[a]pyrene could cause cancers.

Originally developed as an insecticide, the aromatic amine AAF was never put on the market because it was found in 1941 to cause cancer in laboratory studies of many species, including the mouse, rat, hamster, rabbit, and various fowl. And it caused cancer at multiple sites, such as the liver, mammary gland, bladder, lung, eyelid, skin, brain, thyroid, parathyroid, salivary gland, lung, pancreas, gastrointestinal tract, kidney, uterus, renal pelvis, urinary tract, muscle, thymus, spleen, ovary, adrenal, and pituitary. With this rap sheet, the aromatic amine AAF was destined to become a potent model for the study of tumorigenesis.

The importance of the work of the Weisburgers was not confined to the research on AAF but also encompassed the development of standard protocols to test carcinogens. AAF is one of the aromatic amines, which are the same chemicals that were found to cause bladder cancer in the dye industry, and the Weisburgers developed an animal model to predict the relative potencies of chemicals in this family to cause bladder cancers in humans. Fortunately for me, John Weisburger was at the American Health Foundation while I was there, and he was a delightful person with many stories about cancer research at the National Cancer Institute. John had a wonderful full head of white hair and a gentle voice; after entertaining my young daughter Joanna with some stories, she referred to him as the "little bunny rabbit."

The Weisburgers were not the first researchers to explore the role of metabolism in activating chemicals to cause cancer. James and Elizabeth Miller were founding members of the faculty of the McArdle Laboratory for Cancer Research at the University of Wisconsin. They did research on the carcinogenic dye butter yellow (p-dimethylaminoazobenzene), which is somewhat related to the aromatic amines and which had been found to cause liver tumors in rats. In 1948 the Millers found that when butter yellow was administered to rats it produced eighteen known or possible metabolites, one of which, 4-monomethylaminoazobenzene, produced liver tumors. And it did so with the same potency as its parent, butter yellow, identifying it as the carcinogenic metabolite.

David Clayson, at the University of Leeds, in northern England, working with Georgiana Bonser, was investigating the activities of various metabolites of AAF. In 1953, they presented the hypothesis that aromatic amines are converted to ortho-hydroxyl amines during metabolism and that this is what makes them carcinogenic. Eventually the Millers worked out the entire formation of the metabolites of AAF that caused cancer. Another chemical that they explored was the aromatic amine 4-aminobiphenyl, which is found in cigarette smoke and would eventually be linked to the production of bladder cancer. By 1966, the Millers had parsed out the carcinogenicity of several classes of chemicals, including the dyes that had been reported by Rehn in 1895 to cause bladder cancer in German workers. Jerry Wogan at the Massachusetts Institute of Technology discovered that aflatoxin, which is naturally found in mold and is a major cause of human cancer, also required metabolic activation. Other scientists investigated the PAHs found in tobacco smoke and coal tar. All of these agents required metabolism to fulfill their carcinogenic potential. By the mid-1960s, the activation of chemicals by metabolism to carcinogens and reaction with DNA had therefore been worked out for many types of carcinogens. Elizabeth and James Miller coined the phrase "proximate carcinogens" for these activated chemicals.

Although most known human carcinogens require transformation, there are some that don't. Chemicals that interact with nuclear receptors are known as transcription factors and generally do not require metabolism to have their carcinogenic effect. PCBs and chlorinated dioxins are in this category, as are drugs such as phenobarbital, all of which can cause liver tumors in rats. Estrogens are also in this category—and as we will learn later, they can increase the risk of breast and other female cancers in humans.

• • •

The main metabolic system of transforming many of these carcinogens to DNA-reactive proximate carcinogens was an organelle of the liver cell called the smooth endoplasmic reticulum. The interior of a cell has many membranous structures, including the nuclear membrane, separating the nucleus from the cytoplasm, as well as the smooth endoplasmic reticulum and the rough endoplasmic reticulum. The latter is the site of protein synthesis within the cell, and the smooth type is the site of a broad class of xenobiotic metabolizing enzymes generally referred to as P450s. Iron-containing enzymes absorb certain frequencies of visible light that can be detected by spectroscopes. The enzyme cytochrome P450 was named for its strong absorbance at the 450 millimicron wavelength when bound to the lethal gas carbon monoxide.

Further study would reveal that these chemicals were transformed into mutagens that could begin the cancer process with the help of this cytochrome P450.[8] The liver is the metabolic engine of the body and is primarily responsible for the elimination of chemicals from the body. The main purpose of P450 in the liver is to make chemicals more water soluble so that they can be eliminated in the urine. However, the chemical modifications produced by P450 can also produce reactive, cancer-causing carcinogens.

Cytochrome P450 enzymes contain an iron-porphyrin complex that attaches oxygen to chemicals; however, in other parts of the body, similar P450s attach oxygen to chemicals to make hormones. This type of enzyme was first studied in the cortex of the adrenal gland found above each kidney, which is responsible for producing steroid hormones involved in inflammation and the balance of sodium and potassium excretion by the kidney. The raw material for the production of steroid hormones is cholesterol, which is not only ingested as part of our food but also manufactured in the body. It is the building block for many of the body's chemical constituents, including cell membranes and hormones such as testosterone and estradiol. Investigators at Burroughs Wellcome & Co., Inc., in Tuckahoe, New York, were finding that there were two different types of P450 in the liver. One type was increased by treatment of animals with phenobarbital and the other with one of the PAHs, 3-methylcholanthrene.[9] The latter would eventually be cited as the cancer-causing mechanism for a number of chemicals, including PCBs and chlorinated dioxins.

Other important proteins in the body also contain such an iron-porphyrin complex. One is cytochrome oxidase, which is responsible for reacting with oxygen to produce carbon dioxide, water, and ATP, the body's major source of energy. Another is the oxygen-binding protein hemoglobin. Hemoglobin's iron attracts oxygen and carries it from the lungs via the blood to the rest of the body so that it can be reacted with cytochrome oxidase to produce energy. Carbon monoxide also can bind to hemoglobin and prevent it from carrying oxygen, thereby causing asphyxiation.

Enzymes are catalysts: they enable chemical reactions to take place that would not otherwise occur at body temperature. For some reactions, a certain energy barrier must be overcome; the binding of the substrate to the enzyme changes the electrical environment of the molecule and allows the reaction to occur. Enzymes are at work throughout the body. For example, the primary enzyme that metabolizes the common alcohol ethanol in our bodies when we drink is alcohol dehydrogenase, producing acetaldehyde. That, in turn, is rapidly metabolized by another enzyme, aldehyde dehydrogenase, to acetic acid, which is the acid in vinegar. This is easily excreted in the urine.

It is now known that the various cytochrome P450s, using similar mechanisms, are the major pathways for the metabolism of drugs and chemicals encountered by the body. These enzymes are now called the various CYPs rather than P450s, replacing the older term, which has to do with the absorption of light. Of importance for chemicals such as PCBs and dioxins, CYP1A1 and CYP1A2 are induced by these chemicals and are thought to have a role in their ability to produce liver tumors in rodents and melanoma in humans. The interaction of these chemicals with a nuclear receptor called the aryl hydrocarbon receptor, or Ah receptor, also causes changes in many of the cellular regulatory pathways. There are now fifty-five identified and genetically sequenced CYP enzymes that have been found to be involved in various functions, including detoxification of chemicals and pharmaceuticals, fatty acid metabolism, steroid hormone biosynthesis, bile acid metabolism, and vitamin D transformations, among other things. Only a few types are important for cancer development.[10]

11 | Some Carcinogens Directly Affect Genes

I n the late 1950s, P. D. Lawley and P. Brooks, at the Chester Beatty Research Institute, Royal Cancer Hospital, studied whether mustard gas bound to DNA, forming what is called a DNA adduct. They discovered that mustard gas attached to DNA, specifically to a site on guanine.[1] Then in the early 1960s, Lawley and Brookes proposed that the reaction of mustard gas with the specific N7 position of guanine would alter its state of ionization, thereby affecting the specificity of the normal base pairing between guanine and cytosine. They also expanded the same reasoning to alkylation of DNA at the N1 position of either adenine or cytosine, which would prevent the hydrogen bonding of these bases in DNA.

These changes interfere with the faithful replication of DNA via mispairing, which leads to "point mutations" causing changes in the genetic code.[2] The development of point mutations implies a change in the replication of a single base pair of the genetic code. In 1964, Brooks and Lawley published results of studies of the binding of compounds with DNA in mouse skin that corresponded to the carcinogenic activities of these compounds.[3] Then, similar findings were reported by other investigators for several other types of carcinogenic chemicals, including the nitrosamines, AAF, and the azo dyes.[4]

Elizabeth and James Miller had coined the phrase "proximate carcinogens" for the activated chemicals produced by the activity of metabolic enzymes. The chemical nature of the activation had been proposed to be the development of what are called reactive electrophiles, or electron-seeking chemical species, which reacted with the nucleophilic centers in the DNA such as the nitrogens of the purines and pyrimidines. An example of an electrophile is carbonium, which is a positively charged, highly reactive, and unstable type of carbon ion that is part of a molecule such as one of the PAHs found in soot. The positive charge allows the carbon of the chemical to bond with the electron-rich nitrogen of the DNA.[5]

Fred Kadlubar, who had worked with the Millers at the University of Wisconsin, followed up on the observation that several carcinogenic aromatic amines and PAHs reacted covalently with the guanine base of DNA. These produced "bulky" adducts rather than the smaller methyl or ethyl groups of alkylating adducts. To define the actual events that take place during DNA replication, Kadlubar studied space-filling molecular models of DNA containing these adducts for the carcinogens AAF and benzo[a]pyrene. He found that the attachment of these adducts to the base-pairing region of guanine could cause a mispairing of the adducted guanine with another guanine instead of a cytosine, similar to the mispairing that occurs when DNA is affected by mustard gas. This defined the molecular basis for point mutations caused by PAHs and aromatic amines.[6]

The next important breakthrough for toxicology came from a microbiologist at the University of California at Berkeley. Bruce Ames had developed a mutated strain of *Salmonella typhimurium*, the bacteria that causes typhoid fever, which was no longer capable of producing the essential amino acid histidine and therefore required this nutrient in its medium to produce proteins and survive. But if the bacteria suffered another mutation via chemical exposure, one reversing the first mutation, it would again be able to produce the needed histidine. In 1972 Ames collaborated with James Miller to study AAF. They found that the metabolites of AAF were more mutagenic than the parent compound, thereby verifying the need for metabolism to produce the proximate carcinogen.[7] Ames had published a similar article that year regarding PAHs.[8] The next year, Ames modified his testing protocol to include the microsomal fraction of liver homogenate in the incubation, which allowed the metabolism of chemicals to their proximate carcinogens. As a result, eighteen carcinogens, including aflatoxin, benzo[a]pyrene, acetylaminofluorene, benzidine, and dimethylamino-trans-stilbene, were found to be mutagenic. These chemicals were shown to be activated by the added liver homogenates such that the resulting metabolites caused by the liver enzymes formed potent mutagens.[9]

Over the years, the Ames test for mutagenicity became the gold standard for screening for potential genotoxic carcinogens. And with the addition of the liver enzyme activating system, it was able to distinguish between mutagens that required metabolic activation and those that did not. The Ames test would become incredibly important in toxicology because it provided mechanistic information that was relatively easy to obtain. It could screen potential genotoxic carcinogens at a small fraction of the cost of a cancer bioassay in animals.

Bruce Ames had concluded that the mutations caused by carcinogenic chemicals in his bacterial test systems were altering his bacteria's genetic code. His research focused on "frameshift" mutations whereby the interaction of the chemical with DNA base pairing led to an addition or deletion of base pairs in the DNA sequence during replication. This is one type of "point mutation": the alteration in a single base that can lead to a heritable change in the genetic code. The other type of point mutation caused by many carcinogens, such as simple alkylating agents, is a simple base pair substitution caused by mispairing. These were also found to be enhanced by the microsomal fraction of liver homogenate containing P450s.[10]

• • •

Cells have a number of enzymes that can repair the DNA altered by a chemical before a mutation can occur. The first DNA repair enzyme was discovered in 1974 by Thomas Lindahl, of the Francis Crick Institute, which won him the Nobel Prize for chemistry in 2015. After his initial work, many types of DNA repair would be identified, and Lindahl's contribution was the description of "base excision repair"—how a single damaged base could be repaired via removal. Another type called nucleotide excision repair was discovered by Aziz Sancar, of the University of North Carolina; he became the first Turkish scientist to receive a Nobel Prize. The third recipient in 2015 was Paul Modrich, of Duke University, who found the repair enzyme that is called into action when the faithful replication of DNA goes awry. This is not DNA damage caused by a chemical agent but is the "bad luck" mutation that occurs commonly in the stem cells of our various tissues during normal DNA replication. The overall process is called mismatch repair: the wrong base pair is detected, the wrong segment of DNA is cut out and discarded, and then the missing nucleotides are replaced by DNA polymerase. These three repair mechanisms were the first of many to be discovered.[11]

Point mutations are not the only type of changes in the chromosomes that can lead to cancer. The clastogenic effects of some chemicals that cytogenetic analysis has identified as causing wholesale damage to chromosomes can lead to other types of mutations, and these, unlike point mutations, usually cannot be repaired. One such case is the mechanism by which benzene causes acute myelogenous leukemia. Benzene is metabolized in the liver to phenol, hydroquinone, and other hydroxylated products, and in the vast majority of these tests, there were no findings of mutagenicity. Scientists had used two different methods to look for DNA adducts of benzene and its metabolites. One method used radiolabeled benzene and looked for the radioactivity in biochemically isolated DNA fractions. The other used a sophisticated instrument called a mass spectrometer to identify the DNA adducts. However, neither method showed convincing evidence of DNA adducts.

Rather than causing point mutations, benzene and its metabolites cause chromosome breaks and deletions indicative of breaks of both strands of DNA. Benzene requires the metabolism by P450 in the liver, like so many other carcinogens. The hydroxylated metabolites of benzene are responsible for the positive tests for clastogenicity, the type of damage that can be viewed when looking at entire chromosomes. Furthermore, when comparing the results of benzene to other clastogens, the pattern of types of positive tests were most similar to inhibitors of an enzyme called topoisomerase II. Benzene metabolites had also been shown to inhibit topoisomerase II, which is an enzyme crucial to the successful DNA replication process.[12]

When the cell is not involved in the process of replication, each chromosome has its two long strands of DNA bonded to each other in a double-helical pattern. During the early G1 phase of cell division, the cell synthesizes nucleotides so that they will be ready to form the new strands of the chromosomes. At this point in the process the cell also tries to repair any DNA damage that has occurred. In the S phase, the two strands of DNA separate, and the complementary bases bond, forming two new, identical sets of twenty-three pairs of chromosomes. Next, the long strands of DNA are packaged into chromosomes so that they are able to segregate into the two daughter cells during the M phase. Finally, the chromosomes are unpackaged in the daughter cells so that the DNA is accessible to perform its functions.

The long stringy chromosomes in resting cells tend to get tangled and form knots, but the replication of the chromosome can't take place if there are knots. Instead of having to untangle the knots, which would be almost impossible, a special enzyme called topoisomerase II cuts both strands of the DNA, relaxes the supercoiled part, and puts them back together again, thereby untying the knot. However, if something goes wrong in this process, the double strand can break, causing a loss of part of the chromosome. Topoisomerase II is a target of many cancer chemotherapies, including etoposide and doxorubicin (and its derivatives), which are frontline therapies for myriad cancers, including breast cancers and lung cancers. Topoisomerase II poisons cause double-strand breaks, and the ensuing damage can trigger mutations, chromosomal translocations, or other aberrations.[13]

• • •

If repair processes do not prevent or fix mutations, and if these mutations or other DNA damage occur in critical genes regulating the cell replication process, uncontrolled cell division can occur, leading to cancer development. Two types of such critical genes, known as oncogenes and tumor suppressor genes, were discovered in the 1970s. Oncogenes were first discovered in animal cancers caused by viruses in the 1950s, in particular the Rous sarcoma and certain mouse mammary cancers and leukemias. These discoveries at first led to the hypothesis that human cancers were caused by viruses. Although we now know that chemicals, radiation, and other infectious agents can also cause cancer, some viral genes were indeed found to be important in human cancers. However, the question was which came first: the chicken or the egg? It turns out that the viruses had stolen the genes from mammals and used mutated versions of them to produce cancer in these animals. The original version of the gene that could cause cancer was called the proto-oncogene and was found to be a normal part of the mammalian cell's replication machinery. Some of the names of these proto-oncogenes are *src*, *myc*, *ras*, and *jun*. The mutated version, the oncogene, is responsible for turning on cell division and allowing a cell to become cancerous. The mutated *ras* was the first to be found in human cancer; Robert Weinberg, at MIT, a cousin of the famous cigarette-smoking epidemiologist Ernst Wynder, was one of those who isolated the gene.[14]

Another line of experiments led to the discovery of another set of critical cell cycle regulatory genes called tumor suppressor genes. Henry Harris at Oxford University was taking groups of cells in Petri dishes and forcing them to fuse together. When he fused cancer cells with normal cells, the unexpected happened. Instead of the cancer cells dominating the fused cells, making them act like cancer cells, the opposite occurred. They began to behave like normal cells and lacked the ability to produce new tumors. The only possible explanation was that the genes of normal cells could put the brakes on the genes of tumor cells.

These findings eventually led to the discovery of certain genes in humans that were normally responsible for controlling cell division: tumor suppressor genes. When these important genes are mutated, they lose their ability. The first of these specialized genes was related to retinoblastoma and was discovered by the pediatrician Alfred Knudson initially and then by Robert Weinberg, at MIT, and Thaddeus Dryja, at Harvard. Another tumor suppressor gene, *p53*, which was discovered by Arnold Levine, at Princeton University, is mutated in as many as 60 percent of all cancers.[15]

The protein product of the *p53* gene is capable of halting the gene replication process, as a prelude to cell replication. It allows the cell to repair chemical DNA damage before the replication process causes a permanent mutation. Another pathway that a cell with extensive DNA damage can follow is to die via a process called programmed cell death, or "apoptosis."

The *p53* tumor suppressor gene starts this apoptosis, which allows cells to be discarded with the minimum of collateral damage caused by inflammation. However, when the *p53* gene is mutated, the cell replicates without stopping for DNA repair or causing its own death. It keeps replicating and gathering more and more mutations, eventually leading to cancer.[16]

These mutations of proto-oncogenes and tumor suppressor genes would eventually be put in perspective for the development of a cancer. Bert Vogelstein, at the Johns Hopkins University School of Medicine, reported studies of four mutations found in human colon and rectal cancer in 1988. He determined that his results were consistent with a model of colorectal tumorigenesis in which the steps required for the development of cancer often involve the mutational activation of an oncogene coupled with the loss of several tumor suppressor genes. This stepwise process causes normal colon epithelium to become hyperplastic, leading to the formation of benign tumors and finally cancer, which can metastasize to other parts of the body.[17] Colon cancer is one of the more common cancers in men and women, but no known chemical causes in humans have been found. It does appear that diet, especially processed meat consumption, is likely to be a risk factor. Mechanistically, the various steps in this process occur without a known chemical interaction with DNA, at least for now.

12 | Cancer Caused by Irritation

Besides direct effects on the genome, there are other mechanisms by which chemicals can cause cancer in humans. One of the oldest theories of cancer development is chronic inflammation caused by irritation, as espoused by the Prussian physician Rudolf Virchow. He was also the proponent of the cell theory whereby new cells were formed from existing cells. In 1863 he undertook the publishing of an ambitious three-volume book on tumors, *Die krankhaften Geschwuelste*. These volumes were based on lectures given during the winter of 1862/1863, and they presented the hypothesis that the origin of cancer was at sites of chronic inflammation. Some of the tumors that Virchow identified were in fact not true cancers. However, Virchow recognized the difficulties involved in defining a cancerous tumor from noncancerous tumors such as those produced in tuberculosis, syphilis, and leprosy. Today when we describe a lesion as a tumor, we immediately identify it with cancer or at least a benign neoplasm. But the archaic meaning of the word "tumor" is a swelling of any type.[1]

In the teachings of Virchow's time, three factors were thought to lead to the formation of tumors: local causes, predispositions, and dyscrasias. Dyscrasia was the ancient Gallenistic concept involving the structure of elements, qualities, humors, organs, and temperaments. Accordingly, diseases involved imbalances or dyscrasias in the four humors: phlegm, blood, and yellow and black bile. For Virchow local causes trumped all, and local irritation was continually evoked by Virchow in tumor formation.[2] By this hypothesis, irritants cause tissue injury and ensuing inflammation and therefore enhance cell proliferation. He also noted leucocytes in neoplastic tissues, drawing a connection between inflammation and other cancers. He suggested that the "lymphoreticular infiltrate" reflected the origin of cancer at sites of chronic inflammation.[3]

After Virchow's treatise on tumor development, irritation took a back seat to the interest placed on the chromosomes in cancer cells. The question of irritation or inflammation as a cause of cancer was rejected by Virchow's pupil David Paul von Hansemann, who then went on to study the nuclei of cancer cells. However, Hansemann agreed with Virchow and many others in noting that tumors could arise in conjunction with parasitic infestations.[4]

Possibly the first cancer that was found to be caused by irritation was cancer of the bladder following infection by parasites. A causal association between infection by *Schistosoma haematobium* and squamous cell carcinoma of the urinary bladder was first postulated over a century ago, by Goebel in 1905. Schistosomes are trematode worms ("flukes") belonging to the phylum Platyhelminthes. Snails in fresh water serve as the intermediate host, and under certain conditions they release the larval form, "cercaria," of schistosomes. These cercariae can attach to the human skin and penetrate into the bloodstream, entering the venous plexus of the bladder, where they release eggs. Many of the most severe pathological manifestations of schistosomiasis are caused by the physical and immunological responses of the host to the eggs. The site of cancer is preceded by chronic schistosome-related bladder ulcers, usually occurring in individuals with previous heavy infection.[5]

About 15 percent of the current global cancer burden is attributable to infectious agents.[6] Inflammation is a major component of these chronic infections. Moreover, increased risk of malignancy is associated with the chronic inflammation caused by chemical and physical agents.[7] For example, chronic infection caused by human papilloma virus or by hepatitis B and C viruses leads to cervical and hepatocellular carcinoma, respectively. *Helicobacter pylori* colonization in the stomach leads to gastric cancer. Long-standing inflammatory bowel disease has also been linked to colon cancer.

Some other cancers are found to be caused by another type of irritation: excessive consumption of alcohol. As early as 1910, it was observed in Paris, France, that about 80 percent of patients with cancer of the esophagus and upper region of the stomach were alcoholics who drank mainly absinthe.[8] In the first step of traditional recipes for distilled absinthe, wormwood and other dried herbs such as anise and fennel are macerated in ethanol. This process released thujone, the active principle of wormwood and other aromatic compounds. Next, this mixture is distilled to various alcohol contents, typically ranging from 45 to 72 percent. Because absinthe consumption reached excessive and alarming proportions at the turn of the twentieth century, many European governments, as well as the U.S. federal government, banned the icon of *la vie bohème* through several prohibition acts. The primary concerns were the development of various neurologic symptoms and psychosis.[9]

But it turned out that the exotic ingredients of absinthe were only part of the problem. Epidemiological studies have clearly indicated that excessive drinking of alcoholic beverages is causally related to cancer of the esophagus, and there is no indication that the effect is dependent on the type of alcoholic beverage. Esophageal complications that are often diagnosed in alcoholics include esophagitis and a precancerous change called dysplasia of the epithelium, more commonly called Barrett's esophagus. Ethanol can, depending on dose, irritate the oral cavity, esophagus, and gastric mucosa in experimental animals. A consequential increase in cell proliferation in rat esophageal epithelium was found after feeding it ethanol. There is little evidence of a direct genotoxic mechanism, that is, the reaction of ethanol metabolites with DNA, and therefore the mechanistic cause may be more simply attributable to physical inflammation.[10]

Other types of cancers are found in people who consume excessive amounts of alcohol. For example, ethanol causes liver injury, which can lead to cirrhosis of the liver. In liver cirrhosis, the normal cellular structure is replaced with fibrous tissue, which is strongly associated with an increased risk for developing liver cell cancer. It is called cirrhosis because the liver becomes hard like a rock and can be felt during a physical examination. Alcoholic liver cirrhosis is considered to be a precancerous condition.[11] Ethanol-related hepatocellular carcinoma without preexisting cirrhosis is rare.[12]

• • •

The fact that infections and alcohol consumption can cause cancers in people is clearly attributable to large chronic irritation and an overwhelming amount of exposure. Toxicity caused by other chemicals can similarly produce tumors in animals without reacting with DNA or causing any direct damage to DNA. Such chemicals are considered to act via "nongenotoxic" mechanisms, of which several types have been described. Among them are the tumors produced by Berenblum and Shubik via the mechanism called "tumor promotion." They used croton oil, which is very irritating to the skin, to produce what they described as an "epicarcinogenic action." A single application of benzo[a]pyrene, followed by repeated applications of croton oil, was sufficient to lead to tumor production by irritation.[13]

Based on these experiments, the paradigm for chemical-induced cancer in experimental animals involves "initiation" and "promotion." Initiators are usually chemicals that directly react with DNA, such as those that are positive in the Ames assay. On the other hand, tumor promoters can enhance the development of cancer in other ways. Tumor promoters may act to enhance cell proliferation through hormonal effects, such as estrogens in the mammary gland or androgens in the liver. Another tumor-promotion mechanism is the inhibition of apoptosis, allowing a cell to survive that would otherwise die given its DNA damage. There can be several stages whereby normal cells become benign tumors and then cancer. However, in contrast to the sequence of cancer development in humans described by Vogelstein in the previous chapter, chemical-induced tumors in experimental animals rarely metastasize.

• • •

Apoptosis, or programmed cell death, is the body's means of disposing of damaged cells with the least amount of damage to the surrounding tissue. When the damage to a cell is so extensive that the cell cannot provide the basic energy needed to perform apoptosis, it dies by releasing enzymes that digest the cell; this is called "necrosis." The accompanying inflammatory response, involving the invasion of white blood cells and other cells, leads to the formation of scar tissue. Toxicity on a cellular basis can be caused by the reaction of activated chemicals on proteins or membranes rather than on DNA. It is often the result of the inhibition of some metabolic process such as the ability of the cell to produce ATP or other critical enzymes required for energy production. When a chemical causes toxicity of the cell leading to necrotic change, subcellular organelles called lysosomes release enzymes that break down DNA, proteins, and lipids.[14] In either apoptosis or necrosis, the cell dies so it cannot produce a cancer. However, in the tissue's attempt to replace the dead cells, cancer can develop.

This nongenotoxic method of cancer development is based on the notion that the primary effect of toxicity leads to tissue repair through increased cell proliferation.[15] This type of cancer formation is often caused by the very large toxic chemical exposures given during bioassays in experimental animals. For example, the ingestion by an animal of a large toxic dose of a chemical on a daily basis will go immediately to the liver of the animal for detoxification, thereby producing liver cancer. Such toxic events are believed to be a major cause of chemical-induced increases in tumor incidences in some cancer bioassays in which animals are given a "maximum tolerated dose" for their lifetimes. In experimental animals such as rats and mice, the incidence of cancers by such nongenotoxic mechanisms can be quite high, involving half or even almost all of the animals.

Toxicity produced by chemicals often increases the spontaneous rate of tumors in experimental animals, which are often already occurring at high rates. This is clearly the case for liver tumors in rats and certain strains of mice. The toxicity does not produce changes in DNA but instead enhances tumor formation from spontaneously generated precancerous lesions that are prevalent in a high background cancer formation. The toxicity-driven cell proliferation causes cells to ignore requirements for the repair of endogenous DNA damage, which leads to mutations. The *p53* tumor suppressor gene that is responsible for halting cell replication when the cell detects DNA damage can be overwhelmed by toxic effects. When the cell's call for replication given the need to replace damaged cells overrides the ability of *p53* to halt the replication, mutations are produced and can increase the production of cancers.

Nongenotoxic effects of chemicals are temporary and reversible at the beginning, depending upon the continued presence of the chemical. However, these nongenotoxic effects when produced at a high enough level and for a sufficient time can ultimately lead to the types of genetic changes that cause cancer.[16] Although it is now clear that proliferation of cells alone does not cause cancer, sustained cell proliferation in an environment rich in inflammatory cells, growth factors, activated stroma, and DNA-damage-promoting agents promotes the risk of cancer development. Sometimes the insult overwhelms the body's ability to deal with irritation and the resulting inflammatory reaction. Major human diseases caused by silica and asbestos are produced by the irritation these agents produce. Even cigarette smoke causes irritation, which is thought to contribute to its ability to cause lung cancer. The resulting conditions are silicosis, asbestosis, and bronchitis. The cancers produced are lung cancers for all three, along with mesothelioma for asbestos.[17]

• • •

An important step in understanding how inflammation can cause mutations came from studying the effects of white blood cells, which the body recruits to the site of inflammation. In 1981, Sigmund A. Weitzman and Thomas P. Stossel, of the Hematology-Oncology Unit, Massachusetts General Hospital, and Department of Medicine, Harvard Medical School, made an important discovery. They tested human white blood cells, leukocytes, in the Ames test and showed that they produced mutations. Since inflammation involves leukocytes, which are known to produce reactive oxygen species, they hypothesized that inflammation was producing mutations by generating these oxygen radicals. To test their hypothesis, they compared the mutagenicity of neutrophils, which are one type of white blood cells, from a patient with chronic granulomatous disease (CGD) with that of neutrophils from normal individuals. In CGD, neutrophils have a defect in the reactive oxygen-generating system and are thus unable to generate superoxide radicals and hydrogen peroxide. In other respects, they are normal phagocytes. They found that the mutagenicity of cells from the CGD patient was markedly diminished compared to that of normal cells.[18]

It may be surprising that the very substance humans require for life—atmospheric oxygen—would also play the role of a toxic chemical. But life on Earth began without oxygen, in an anaerobic environment housing an array of anaerobic organisms. Oxygen was first produced by one of these—the blue-green algae—in a photosynthetic process. The increasing level of oxygen in the atmosphere was toxic to anaerobic organisms, which had to cope with its introduction into their environment either by retreating into places where this reactive molecule could not reach them, such as deep geothermal vents, or by adapting to it. Some organisms adapted by evolving antioxidant defense mechanisms so that their proteins, lipids, and DNA would not be destroyed by oxygen. And they managed to utilize the reactive nature of oxygen to produce energy, stored as ATP, thereby introducing the animal kingdom.

We all know that at temperatures where things burn, oxygen reacts with carbon to form carbon dioxide. Oxygen by itself is not very reactive at body temperature, even though technically it is considered a free radical, but through various mechanisms, oxygen can be transformed into a number of reactive oxygen species (ROS), which are much more reactive free radicals. Because of their reactive nature, these ROS combine quickly with other atoms and molecules and therefore do not exist for a very long time. Mostly they will dissipate harmlessly or be removed by reaction with the protective antioxidants in the body. The main antioxidant in the body, glutathione, is a polypeptide made up of three amino acids. Other antioxidants that react with ROS or that bolster the glutathione levels are the vitamins A, C, and E, as well as some enzymes and certain metabolic breakdown products like bilirubin.

If these ROS are not removed, they may react with DNA or other vital components of cells. Estimates have been made that each cell in the human body is normally subjected to DNA-damaging oxidative reactions ten thousand times per day.[19] Almost all of these will be repaired, but a few of these oxidized bases of DNA will mispair during replication and cause mutations. Such free radical reactions are believed to be a source of many human diseases, including cancer. The continual damage caused by these free radicals is also believed to be involved in the aging process. Fortunately, there are enzymes that repair this type of DNA damage; otherwise our cells would be riddled with genetic errors.

13 | Cigarette Smoking

Black, Tarry Lungs

T he year is 1948. A medical student by the name of Ernst Wynder is attending the autopsy of a forty-two-year-old man who had died of lung cancer. When he opens the cadaver's chest, Wynder is aghast at the condition of the lungs: a mass of black, tarry growth. Wynder knows that the patient was a heavy smoker—reportedly two packs a day—and Wynder immediately suspects that his lung cancer was caused by cigarettes.[1]

In hindsight, Wynder's linking of cigarette smoking to black, tarry lungs appears stunningly obvious, but in 1948 it was the kind of intuitive leap whereby great discoveries in science are made. Smoking was so prevalent at the time, at least among men, and so ingrained in postwar American culture that it was effectively invisible to the medical profession. The smoking rate for a forty-two-year-old man in 1950 was 67 percent; thus it would have been more likely than not that Wynder would find black, tarry lungs in his cadaver, whether or not the patient had developed lung cancer.[2]

At that time, several investigators had published articles suggesting that cigarettes might be important in the production of lung cancer, but large-scale clinical studies were lacking. Most physicians ignored these early studies. Wynder and his parents had fled Germany when he was an adolescent. His fluency in German led him to a German publication by F. H. Muller, which he found was a careful but limited clinical study offering good evidence that heavy smoking was an important cause of lung cancer.[3]

Wynder had gotten a bee in his bonnet, and he was determined to test the hypothesis that cigarettes caused lung cancer. He was on his surgical rotation at Washington University in St. Louis, and he approached the chief of the department, Dr. Evarts Graham, about doing a study based on his observation. Graham was a renowned thoracic surgeon who had helped pioneer the surgical treatment of cancer of the lungs. He was also a cigarette smoker. He agreed to sponsor Wynder's study even though he disagreed with the notion that cigarettes caused lung cancer.[4] Apparently, he wanted to teach this upstart medical student a lesson.

It is one thing to make a correlation between a factor like cigarette smoking and a health effect like lung cancer; it is another to prove a causal connection. Wynder may have had the advantage of looking at the problem with fresh eyes, but the tools he would need to explore it had not yet been developed. Surgeons like Graham had made key advances in the surgical treatment of cancer, but their efforts shed no light on its origins. The things Wynder needed were advances in epidemiology, the science of teasing out patterns of disease in populations.

Wynder and Graham faced several scientific challenges in planning their study. The first was to choose the most appropriate study design. The most straightforward approach would have been to conduct what is called a prospective study, in which a group (or cohort) of smokers is compared with a cohort of nonsmokers over time to see whether the smokers exhibit a higher rate of lung cancer. Prospective studies yield good data, but they take time. One needs to keep track of the cohorts for years before enough cancer cases develop so that a difference between the groups can be detected.

Wynder didn't want to take that much time, so he decided to do what is called a case-control study. He took two groups of patients—those admitted to the hospital with lung cancer (cases) and those admitted without for other reasons (controls)—and looked for differences in their smoking habits. If smoking were more prevalent among the lung cancer patients than the controls, or if the cancer patients tended to be heavier smokers or to have smoked longer than the controls, a causative link between cigarette smoking and lung cancer might be established. The advantage of this type of study is that it is retrospective; the cases and controls have to be contacted only once.

In 1948, the case-control study method was not yet a well-established technique in epidemiology. In choosing it, Wynder and Graham were running the risk that their results would not be considered very reliable. The problem in most such retrospective studies is that the only way to ascertain relevant information from the patient's past is to ask them. A patient who knows he has cancer is naturally going to want to blame it on something, and so he may over-report past exposures. Or, if the suspected factor is behavioral, like smoking, he may under-report it, hoping to blame the illness on something outside his control. These are called recall biases, and even the interviewers are subject to it. It is natural for an investigator to look for information that will confirm his hypothesis and so distort the patient interview in subtle ways. Wynder tried to minimize the effect of recall bias on his data by developing a standardized, detailed questionnaire for the interviewers to follow.[5]

It is also important in any epidemiological study that the cases and controls be as well matched as possible in all factors other than those under study; they should be as close as possible to one another in age, sex, geography, and so forth. It is also important that there is an accurate definition of the disease under study. In Wynder and Graham's study, lung cancer cases were verified by microscopic confirmation, and the controls were random patients admitted to the participating hospitals matched for age and sex. Wynder and Graham also did a separate analysis with controls that had lung diseases other than cancer.

Wynder and Graham did indeed find a correlation between cigarette smoking and the incidence of lung cancer in their study that was deemed strong enough to be considered causal; not only was it related to smoking but also to the amount of smoking. They announced their preliminary findings at the National Cancer Conference in Memphis in February 1949. Published in 1950 in the *Journal of the American Medical Association*, the study, entitled "Tobacco Smoking as a Possible Etiologic Factor in

Bronchiogenic Carcinoma: A Study of 684 Proven Cases," is now considered to be a landmark paper both in epidemiology and toxicology. In epidemiology it had provided one of the first examples of the use of the case-control methodology. In toxicology, it identified cigarette smoke as the most important—and preventable—cause of human cancer.[6]

Evarts Graham was so impressed by the results of their study that he quit smoking in 1951, but he was under no illusion that others would be receptive to their message. Wynder related what Graham said to him when he reviewed the final manuscript of their paper.

> He said, "you are going to have many difficulties. The smokers will not like your message. The tobacco interests will be vigorously opposed. The media and the government will be loath to support these findings. But," he added, "you have one factor in your favor." Expectantly, I got off my chair to hear what would follow. "What you have going for you," he exclaimed, "is that you are right."[7]

Graham died of lung cancer in 1957.[8]

A few months after Wynder and Graham's paper appeared, Richard Doll and Austin Bradford Hill published another case-control lung cancer study in England showing similar findings.[9] Twenty London hospitals were asked to cooperate by notifying all patients admitted to them with carcinoma of the lung, stomach, colon, or rectum. What Doll and Hill described as an almoner (a hospital social worker) was dispatched to interview each patient and obtain his smoking history. There were 1,732 such cancer patients, and 743 general medical and surgical patients served as controls. A very high rate of smoking was found in all the patients, with or without cancer, making the study somewhat difficult. But, using sophisticated statistical methods, Doll and Hill found that only 0.3 percent of men with lung cancer were nonsmokers, whereas 4.2 percent of controls were nonsmokers. Also, twice as many lung cancer patients were found to be heavy smokers at the time of their diagnosis compared to controls. Their case-control study results were not as robust as those of Wynder and Graham because of the high percentage of smokers in both cases and controls, but the findings contributed additional data to demonstrate causation.

Doll and Hill next published a prospective study of British doctors. At the end of 1951, some forty thousand men and women on the British Medical Register replied to a questionnaire about smoking habits. On that basis they were divided into nonsmokers and three groups of smokers (including ex-smokers), according to the amount they smoked at that time. The preliminary report of this study indicated that there was an increase in lung cancer deaths among smokers.[10] When they published their results after five more years of study in 1956, the death rate of the heavy smokers from lung cancer was twenty times the death rate of the nonsmokers.[11] This study was particularly powerful because it was not a case-control study but a prospective study. It followed two cohorts through a period of time without knowing what their outcomes would be; thus it was less prone to recall bias. This study also put aside the argument that occupational factors, including industrial exposures, could have confused the findings with cigarettes, since all of the study participants were doctors.[12]

• • •

The claim that cigarette smoking caused lung cancer was strenuously opposed by the tobacco industry, and these early studies did not have the impact they should have had. By the time Wynder's paper had come out, he had moved from St. Louis to the Sloan Kettering Institute (now Memorial Sloan Kettering Cancer Center), in New York City. Phillip Morris tried to co-opt and suppress Wynder's work by giving donations to Sloan Kettering and then using that influence to keep Wynder from speaking on the subject. According to Fields and Chapman, in the fall of 1962, Dr. Horsfall and other Sloan-Kettering officials began subjecting Wynder to more rigorous screening procedures before letting him speak in the name of the institute.[13] Wynder was also incentivized to study cigarettes, with tobacco industry funding, to find a safer cigarette that would minimize the lung cancer risk. He continued this work for many years, even after he left Sloan Kettering, because he understood the practical difficulties of trying to get the public to stop smoking.

None of this did anything to shake Wynder's belief that cigarette smoking caused lung cancer, and he began to search for a plausible mechanism.

Doll and Hill had postulated that it was the arsenic in tobacco smoke that caused lung cancer. The chemical additives in cigarettes were a closely guarded industry secret. In 1953, Wynder along with Adele Croninger produced concentrated tars with a cigarette smoking machine. They shaved the hair off mice regularly before applying the tars to their backs, slowly increasing the amount of tars so that the substances wouldn't just kill the mice outright at the outset. The mice rapidly grew tolerant to the toxic effects of the tars so that more concentrated solutions could be given. After the dissolved tar was applied to the backs of mice three times per week for about one year, about half developed skin cancers. Wynder still didn't know exactly how cigarette tar was capable of producing cancer, but he and his collaborators had produced one of the first experimental studies to support the claim that cigarettes cause cancer in humans.[14]

Published in *Cancer Research* in 1953 under the title "Experimental Production of Carcinoma with Cigarette Tar," Wynder's was one of five papers that reported their experimental results using cigarette tar condensate obtained with a smoking machine. In another study, different strains of mice were used. In yet another, the ears of rabbits developed cancer from the application of cigarette tar. Tars produced from pipes and cigars had a similar effect of producing skin cancer after prolonged application.

But even this experimental evidence would come under attack by Clarence Cook Little, who had introduced Wynder during his medical school years to experimental carcinogenesis at the Bar Harbor lab. Little was about to retire and had received a position as scientific director of the Tobacco Industry Research Council in 1954, founded in response to Wynder's mouse experiments. Wynder's research was attacked by Little, who was committed to the viral theory of cancer. He questioned why experimental animals rarely came down with lung cancer from cigarette smoke. According to Little, the greater rates of lung cancer in smokers were correlated but did not provide evidence of causation.[15]

Eventually in 1956, the National Cancer Institute, the National Heart Institute, the American Cancer Society, and the American Heart Institute jointly reviewed the available studies and concluded that cigarette smoking caused lung cancer. Based upon these findings, Surgeon General Leroy E. Burney, who was appointed by President Eisenhower, issued statements in 1957 and 1959 claiming that cigarette smoking was a causative agent for lung cancer.

A much stronger statement on the matter would come in 1964, when Surgeon General Luther L. Terry, a Kennedy appointee, formed an advisory committee to review the data. Their report, "Smoking and Health,"[16] concluded that cigarette smoking was associated with a 70 percent increase in death rates in males and a lesser increase in females. For lung cancer, there was estimated to be a twenty-fold increase in risk among heavy smokers, and cigarette smoking was found to be the most important cause of chronic bronchitis. The report also concluded that cigarette smoking was a more important cause of chronic lung disease than atmospheric pollution or occupational exposures. Higher mortality rates from cardiovascular diseases were also found to be associated with smoking.[17]

Austin Bradford Hill and Richard Doll were knighted in 1961 and 1971, respectively, for their contributions to public health, but in the United States Wynder continued to struggle against the tobacco industry–funded headwind at Sloan Kettering. Eventually he left and founded the American Health Foundation in 1969, and his disease prevention research institute received generous funding from the National Cancer Institute. The main focus of the American Health Foundation was the identification of the exogenous causes of disease, including tobacco, diet, and chemicals. In an ironic twist, Wynder also continued to receive funding from his old nemesis, the tobacco industry, which by this time was making efforts to reduce the amount of cancer-causing constituents in their cigarettes. This allowed him to continue his efforts to identify the actual chemical constituents in cigarette smoke that caused cancer.

Wynder was an outspoken leader in preventive health care. Even though the verdict on cigarette smoking and human health had come in resoundingly in 1964, there had been little action in implementing broad preventative health measures on cigarette smoking. In 1975 he organized a symposium at Rockefeller University titled "The Illusion of Immortality" in order to discuss the matter. The symposium included some of the leading intellectuals of the day, including Ashley Montagu, the author of *The Natural Superiority of Women*; the existential psychologist Rollo May; and the theologian William Sloan Coffin. Also present was the famous heart surgeon Michael DeBakey, who was a pioneer of open-heart surgery and described how his recovering patients often resumed smoking after surgery. Wynder gave the summary for the conference, saying, "Man tends to live for the moment, believing that tomorrow will take care of itself. He does not like to consider the possibility of crippling illness or death. . . . A man doesn't take his car for granted. He brings it in for a regular checkup. But he takes his body for granted."[18]

• • •

Ernst Wynder had a methodology that he developed during his research on cigarette smoking, and he modeled the various divisions of the American Health Foundation on this approach. He was first and foremost an epidemiologist, believing that important clues for the causes of human disease could be found by looking for patterns of disease in various human populations. So in his foundation he established a Division of Epidemiology and installed Steven Stellman as his primary epidemiologist. According to his method, the results of epidemiology studies identified presumptive causative factors for diseases. Then, because laboratory studies were needed to provide what we now call biological plausibility for the associations epidemiology turned up, the foundation was equipped to do animal experiments. According to Stellman, the juxtaposition of laboratory scientists with epidemiology created a synergy that promoted new ideas for epidemiology.[19]

Next, additional epidemiology studies would follow up on the laboratory studies to determine whether alterations could be made in the nature of the causative agent. In the example of cigarettes, alterations in the composition of the tobacco or filtration of smoke were studied in smokers to see if lung cancer rates could be decreased. In effect, Wynder had anticipated the e-cigarette, which delivers the nicotine to the smoker without the tars that cause lung cancer.

Studies performed at the American Health Foundation found some surprising results. In a study published in 1989, Wynder looked at the question of whether smoking filter cigarettes would reduce the risk of lung cancer. Again, he used the case-control study design. The cases and controls in this study were interviewed between 1969 and 1984, in twenty hospitals in nine

U.S. cities, as part of a hospital-based ongoing study of smoking-related cancers. They found that people increased the number of cigarettes that they smoked each day after switching to filter cigarettes and that this was associated with a higher incidence of lung cancer in those patients. In other words, filters made things worse.[20]

Wynder reported that the risk of squamous cell carcinoma, which was the type of lung cancer that he had originally studied, was reduced among the relatively few smokers who only smoked filter cigarettes during their lifetimes. But smokers of low-tar filtered cigarettes were found to smoke more and that their puff volume was greater. That meant that the deeper parts of the lung were being exposed with each puff, which included cells of the terminal bronchioles and alveoli, where exchange of oxygen and carbon dioxide occurs between the air and the blood. The shallower part of the lung is the bronchi, and there, in smokers, the normal cellular lining would first transform into squamous cells, which were like skin cells, and these would then be become squamous cancer cells. In the deeper part of the lungs, the cells did not become squamous first but were directly transformed into cancer of the adenomatous type. So filter cigarette smoking led to an increase in adenocarcinoma of the lungs.

Using this information, the epidemiologists could take another look at their studies and try to make sense out of the importance of their findings. This was Ernst Wynder's paradigm for carcinogen research. First, take clues from epidemiology studies, especially cross-cultural studies. Next, study these chemicals or lifestyle factors in the laboratory and try to determine the mechanism by which they cause cancer. Then, evaluate the importance of the epidemiology studies based on the animal studies by understanding the plausibility of their findings. Finally, develop more epidemiology and laboratory studies to investigate further the causal relationships and mechanisms.

All of this exciting research by Wynder and his associates demonstrates the importance of the collaboration between epidemiology and toxicology for figuring out what causes cancer and how to prevent it. It was a very heady time for toxicology, using the methods that had been developed over the 1970s and 1980s to test for cancer. Through studies in humans and in experimental animals, toxicologists had learned a tremendous amount about human diseases caused by various chemical agents.

14 | What Causes Cancer?

S everal scientists in the 1970s attempted to determine the extent of preventable cancers caused by environmental agents. Ernst Wynder, from the American Health Foundation, and Gio Gori, of the National Cancer Institute, reviewed some of these estimates in 1977, finding that 90 percent of cancers might be caused by environmental factors, including chemicals, cigarette smoking, and nutrition. This included occupational exposures, which accounted for an estimated 1 to 10 percent of associated cancers.[1] Although the word "environmental" has become synonymous with chemicals in the ambient surroundings, these investigators meant factors other than those that were genetic or unknown. Another way of looking at this is that some cancers could be avoided by preventing exposures from outside the body—thus the term "xenobiotics"—as opposed to those cancers that are produced by the normal processes in the body that might cause genetic errors or other effects that could lead to cancer.

In 1978 the National Institutes of Health, Environmental Health Sciences, and Occupational Safety and Health Administration submitted a report to the Occupational Safety and

Health Administration claiming that a minimum of 20 percent of all cancer deaths in the United States were caused by occupational exposures. The OSHA report said that the largest risks were from asbestos exposure. In the same timeframe, Congress requested a comprehensive report by two famous epidemiologists, Richard Doll and Richard Peto, the same British investigators that had published one of the definitive reports of cigarette smoking and lung cancer in 1950. This commissioned report was aimed at informing Americans about the preventable causes of cancer. Their findings were a watershed in the field of toxicology because they reported a startling fact: industrial chemicals in the workplace and the environment were not among the major preventable causes of cancer and contributed a much smaller percentage of overall cancer incidence. The report Doll and Peto produced was well documented. They concluded that the OSHA reported much higher estimates out of political rather than scientific considerations.[2]

So what are the estimates of avoidable cancer that these scientists found, and how did they make their calculations? Cancer is a difficult disease to study given its latency, the delay between exposure and the production of disease. Fortunately, in studying populations, these epidemiologists developed clever ways of providing good estimates of the various types of cancer in different countries. They then compared the differences. The difference between the country with the lowest rate for a cancer was compared to the country with the highest rate, and that difference was considered to be the quantifiably avoidable excess amount. They also investigated rate changes when populations migrated and over time in stable communities.

Next, causative agents were examined to account for this excessive amount of cancer. Using other epidemiology studies, such as lung cancer rates for smokers and nonsmokers, the amount of lung cancer caused by cigarette smoking was calculated. For occupational exposures, the same was done for studies of different chemical exposures and then summed up. Although these estimates were made from statistics available in the late 1970s, they are still relevant; they were confirmed in a 2001 publication by Henry Pitot and Yvonne Dragan. [3] This estimate also was generally supported by an analysis by Blot and Tarone in 2015.[4]

Of most importance to toxicologists, Doll and Peto attributed only about a 1 to 2 percent contribution to the past cancer burden from environmental pollution and another 2 to 4 percent from occupational exposures.[5] Another contributing issue in toxicology is drugs. One change over time is that Doll and Peto estimated that medicines and medical procedures had accounted for about 1 percent of human cancers at that time. Most of these were attributed to therapeutic ionizing radiation, estrogens, and oral contraceptives.[6] Today we can add cancer chemotherapeutic agents to this list because they can cause secondary cancers, especially leukemia, in their wake.

One major exogenous source of cancer identified by Doll and Peto was tobacco. During the past century, 30 percent of all preventable cancer was estimated to be caused by cigarette smoking. Smoking cessation programs, legal actions against tobacco companies, and bans on public smoking provide hope that cancer can be reduced by one-third though eliminating this cause. Clearly the outstanding research and courage of scientists who tackled cigarette smoking has made the most important contribution to the prevention of cancer. And the rates of lung cancer are now decreasing thanks to their efforts. Some of this decreasing risk of lung cancer could also be attributed to controls of air pollution and occupational exposures from asbestos. Some additional small contribution to the lung cancer decrease may be attributable to controls of other air pollutants eliminating from smelters, power plants, and diesel exhaust.

According to studies by Wynder and Doll, cross-cultural epidemiology studies identified diet as perhaps the most important cause of cancer. Diet was assigned 35 percent of preventable cancer deaths based on comparative epidemiology studies, although this quantification was characterized as highly speculative. It is still difficult to know the exact components of a person's diet that can produce cancer. Comparing different population groups, such as Japanese and Americans, Finns and other Europeans, or Seventh-Day Adventists and other Americans, Wynder identified fat, low fiber, and meat in our diets as major causes of cancer. Cross-cultural studies suggested that cancer could be prevented by increasing fiber, decreasing fat, and increasing the consumption of fruits and vegetables containing calcium and the vitamins E, C, and beta carotene.[7]

Since these first attempts to link cancer to differences in diet, there have been numerous studies in nutritional epidemiology. The review by Blot and Tarone in 2015 summarized the state of knowledge by echoing the characterization by Doll and Peto: "a chronic source of frustration and excitement to epidemiologists." For example, fat consumption has not been proven to be a risk factor for breast cancer, which was one of Ernst Wynder's favorite research topics. The only food-related risk for cancer that has been established is processed meat consumption and colon cancer.

Besides assigning risks to chemical contaminants in food or to dietary components such as fat or lack of fiber, overnutrition was also cited as a major factor. Obesity had been well studied experimentally and found to increase the incidence of tumors. As noted by Henry Pitot and Yvonne Dragan at the McArdle Institute at the University of Wisconsin, rodents allowed to eat as much as they want will have four to six times the numbers of cancers compared to those who have their caloric intake restricted.[8]

The number of human cancer cases caused by obesity is now estimated to be 20 percent of all cancers in the United States, with the strongest evidence for an association of obesity with the following cancer types: esophageal, stomach, colon and rectum, liver, gallbladder, pancreatic, postmenopausal breast, uterine, ovarian, kidney, thyroid, multiple myeloma, and one type of brain cancer, meningioma.[9] Risk of developing endometrial cancer is four times greater for obese versus normal-weight women, and this results in 50 percent of all endometrial cancer in the United States being attributable to being overweight. Obesity also accounts for 35 percent of esophageal adenocarcinomas, 15 percent of colorectal cancer, 17 percent of postmenopausal breast cancer, and 24 percent of kidney cancer.[10]

Obesity is associated with metabolic and endocrine abnormalities, including changes in sex hormone metabolism, insulin and insulin-like growth factor (IGF) signaling, and adipokines, or inflammatory pathways. According to the IARC, evidence for obesity-induced sex hormone metabolism and chronic inflammation in cancer risk is strong.[11] These effects of obesity can lead to increased cell proliferation in many tissues.

Another avoidable risk of cancer is infection, particularly by viruses, which has been shown to be a major cause of cancers. Papilloma virus produces carcinoma of the cervix, vulva, vagina, penis, anus, oral cavity, oropharynx, and tonsils. Hepatitis viruses cause liver cancer and non-Hodgkin's lymphoma. Human immunodeficiency virus (HIV-1) has been strongly associated with Kaposi's sarcoma, non-Hodgkin's lymphoma,

Hodgkin's lymphoma, and cancers of the cervix, anus, and conjunctiva. Bacterial infection with *Helicobacter pylori* causes stomach cancer, and the parasite *Schistosoma haematobium* causes bladder cancer.[12]

Another risk factor known for non-Hodgkin's lymphoma is immuno-suppression from drugs used to prevent organ transplant rejection. An intriguing Australian study of increased cancer rates associated with immunosuppressive therapy published in 2006 found that besides cancers with a known viral cause, cancers of the salivary gland, esophagus, colon, gallbladder, lung, uterus, and thyroid, as well as leukemias, were of higher incidence in patients treated with immunosuppressive therapy after receiving kidney transplants. In contrast, other common cancers that are thought to be primarily hormone driven were not at increased incidences in these patients. This study seems to implicate infections or a lack of immune defenses in these additional cancers.[13]

Doll and Peto found that geophysical factors accounted for only about 3 percent of all cancer deaths, with ionizing background radiation and UV light each contributing about half to this risk. Ultraviolet light contributes to malignant melanoma, a fatal form of skin cancer. Ionizing radiation has been found to cause numerous cancer types because this radiation can cause chromosomal damage in any tissue or organ. However, since nonmelanoma skin cancers (squamous and basal cell) are by far the most common forms of skin cancer, we have to be careful of relying only on mortality statistics because these other skin cancers are rarely fatal. Doll and Peto estimated that more than 80 percent of these types of skin cancers were attributable to UV radiation exposures.

• • •

Rather than looking at causative factors for cancers, another method is to examine different cancer types and see whether risk factors have been found. For some of the common cancers, such as breast, bladder, and the leukemias and non-Hodgkin's lymphomas, there have been no significant changes in frequency during the past twenty years, even as industrial pollution has been reduced. This is also true for some of the more infrequent cancers, such as some brain, endometrial, oral cavity, and testicular cancers;

conversely, the numbers of melanomas and thyroid and liver cancers have been increasing. Therefore, the observation that decreases in pollution in our environment have not been followed by changes in the rates of these cancers indicates that there are probably other more important causes than environmental ones involving chemical pollution. The numbers of colon, prostate, stomach, ovarian, and cervical cancers have been going down, but many of these decreases can be attributed to early detection of precancerous lesions by colonoscopy or Pap smears or, in the case of cervical cancer, the vaccine for papilloma virus.

Prostate cancer is the second most commonly occurring cancer in men after skin cancer, yet it is the most common cancer in men for which toxicologists have found no chemical cause. Identified risk factors are age, ethnicity, and family history. There is also some association with obesity. Colon cancer is the third most common cancer in men and women. Major risk factors are age, obesity, family history, prior inflammatory bowel disease, and the consumption of processed meat.

We still haven't identified any industrial causes of breast cancer. Earlier in this book, we examined the work of Bernardo Ramazzini, the eighteenth-century physician and father of occupational medicine, who described the diseases of miners and other occupations. Perhaps one of Ramazzini's most notable observations occurs in his chapter "Diseases of Wet-Nurses." This extensive chapter in his treatise includes a discussion of the relationship between the uterus and breast, in the context of lactation. Ramazzini reports that "we must admit this sympathy between the breasts and the uterus: for experience proves that as a consequence of disturbances in the uterus, cancerous tumors are very often generated in women's breasts, and tumors of this sort are found in nuns more than in any other women." He noted that every city in Italy has several religious communities of nuns and that one could seldom find a convent that does not harbor this accursed pest, cancer, within its walls.[14]

Breast cancer is correlated with factors that expose women to either endogenous or exogenous sources of estrogens. In addition to physician-administered estrogen, most of the established endogenous risk factors for breast cancer seem to operate through hormonal pathways. This is evident from the established associations of increased breast cancer risk with early

age at menarche, late age at first full-time pregnancy, and late age at menopause, all of which add to the total amount of estrogen exposure during a
woman's lifetime.[15] Brian MacMahon and Phillip Cole, from the Epidemiology Department at Harvard's School of Public Health, and four other
international investigators reported a collaborative study of breast cancer
and reproductive experience from seven world regions. Women having their
first child at eighteen years of age or less had only about one-third the breast
cancer risk of those whose first birth was delayed until the age of thirty-
five years or more.[16] Other factors, such as the number of pregnancies, have
also been shown to decrease breast cancer risk. A study in India reported
that women who had three or more pregnancies lowered their risk of breast
cancer by at least half.[17] These findings have led to the hypothesis that the
modern rise of breast cancer risk is based on two main factors: women are
living longer and are having fewer children. Breast cancer risk has been
related to the total number of menstrual cycles that women experience.
Fewer total cycles could be caused by late menarche, early menopause, a
greater number of pregnancies, and the length of breast feeding.[18] Another
factor that increases exposures to estrogen is related to excess body weight.[19]
The only chemicals that have been shown to increase the risk of breast cancer are in hormone replacement therapy and birth control pills.[20]

It has been proposed that industrial chemicals that are estrogenic could
also contribute, but this has never been proven. It turns out that humans
are exposed to large amounts of estrogenic compounds naturally occurring
in food, which may drown out any possible contribution from industrial
chemicals. But these are weakly estrogenic chemicals at best, and by occu-
pying estrogen receptors and not activating them, they are probably anti-
estrogenic if anything.

The case could be made that our myopic vision of the causes of cancer
has distorted our judgment about research priorities. One need only look
at the example of *Helicobacter pylori* infection and stomach cancer. Despite
extensive research, not one industrial chemical has been identified as a
stomach carcinogen in humans. Yet there was incredible resistance to the
notion that bacterial infection could cause stomach ulcers and cancer.[21] The
American Cancer Society now estimates that 63 percent of stomach cancers are caused by this infection. In 1900, stomach cancer was the most

common cancer in the United States. During the twentieth century it dramatically decreased. The decline of stomach cancer in the United States has been attributed to the increase in refrigeration and decrease in the consumption of salted or smoked products.[22] Stomach cancer is still the fourth most prevalent cancer in the world, with the highest number of cases in China, Japan, Russia, and the western coastal countries of South America. Much of this has been attributed to dietary intakes of salted food, which in these regions is used for preservation.[23]

Stomach cancer was intensively studied for chemical causes in foods; various food additives were studied in particular because it made sense to do so. Industrial chemicals in the form of artificial flavors, colors, and antioxidants to prevent spoilage were added to food, and they went directly to the stomach. It made sense that they would cause cancer, especially when they were shown to cause stomach cancer in rats. But there are problems with this hypothesis. First, stomach cancer had been declining remarkably over the twentieth century, while the use of these additives has been increasing. Second, no chemical has been identified as causing stomach cancer in humans. Third, the chemicals shown to cause stomach cancer in rats cause the cancer mostly in the forestomach, a part of the rodent stomach that humans lack.

We now know that the major known causes of liver cancer in humans are hepatitis infections, alcoholism, and aflatoxins that are produced by the molds *Aspergillus flavus* and *Aspergillus parasiticus* in stored foods such as peanuts and corn in warm climates. Liver cancer caused by a combination of aflatoxin and hepatitis is a major problem in China.[24]

Less common but very deadly are brain tumors. There are three main types of malignant tumors in the brain; they have their peak incidences at three different times of life. One occurs in children, another in the elderly, and yet another primarily in middle-aged men. In this latter case, men who are in the prime of their life suffer from an attack on their brain that makes them miserable, taking away their ability to function and eventually their life. An early event in the formation of these brain tumors is a mutation of the *p53* tumor suppressor gene. In most cases where the *p53* gene is mutated, it is a later event in the progression of the cancer. Substantial research effort has attempted to identify chemicals that might cause brain cancer, and again we have come up empty handed.

One form of leukemia, acute myelogenous leukemia, has been clearly linked to substantial exposures to benzene, even though most cases of leukemia are not caused by benzene. Several other industrial chemicals have been reported to cause leukemia or lymphomas. But in spite of these identifications and occupational controls, the rate of leukemia incidence has remained steady during the past twenty years.[25]

• • •

One of the earliest cancer risk factors identified by epidemiologists was age. This relationship of cancer to age is the reason that cancer statistics used for comparison in epidemiology are age adjusted. The Surveillance, Epidemiology, and End Results (SEER) program of the National Cancer Institute tracks information about cancer incidence and mortality based on age, cancer types, geography, and other factors. Based on their data for all invasive cancers combined in 1992 through 1996, 80 percent occurred in people fifty-five years of age and older. The incidence and mortality of cancer rates increased after the age of ten and rose in an exponential fashion until after seventy-five years of age, where the incidence begins to level off but mortality continues to climb.[26]

What does the age relationship tell us about the causes of cancer? Perhaps older people are less capable of fighting off cancer cells, making cancer easier to develop. Perhaps carcinogens accumulate in the body with age. However, from a standpoint of the biology of cancer development, the most plausible mechanism is the accumulation with age of the mutations required to produce a cancer. Cancer is known to be caused by a multistage process, usually requiring several mutations in order to develop.

Bert Vogelstein was at the Department of Pathology, Johns Hopkins University School of Medicine in 1988 and reported studies of four mutations found in human colon and rectal cancer. He determined that his results were consistent with a model of colorectal tumorigenesis in which the steps required for the development of cancer often involve the mutational activation of an oncogene coupled with the loss of several genes that normally suppress tumorigenesis. This series of steps requires time and advancing age.[27]

In 2015 Vogelstein, now at the Department of Oncology at Johns Hopkins, along with Cristian Tomasetti, who had joined the department's Division of Biostatistics and Bioinformatics in 2013, ventured to determine how much cancer is a result of normal physiological processes in the body that require time. They did this by comparing the rate of divisions of stem cells and the incidence of cancer in different cell types of organs. Stem cells are the dividing cells in tissues that replenish the deaths of mature cells so that their numbers remain relatively constant. For example, in the skin the basal layer contains the stem cells, and as the cells mature, they travel closer to the skin surface and eventually are sloughed off. In the colon, there are cells in a certain part of the intestinal mucosa that divide in order to replenish the mucosal cells that are sloughed off into the feces.

Tomasetti and Vogelstein estimated the number of cell-type-specific stem cells in an organ that may become cancerous and multiplied them by the numbers of times these cells divide. Next, they did a statistical test for correlation of this product with the cancer risk for these cells. Their finding was astonishing and controversial: they proposed that for two-thirds of cancer, the difference in cancer rates between organs or tissues could be explained by the "bad luck" of spontaneous mutations acquired over time occurring simply from cell division.[28]

This proposition caused an outcry from some in the cancer research community, especially the toxicologists who were invested in cancer prevention efforts involving chemical exposures. They interpreted these findings to indicate that only one-third of cancer could be attributed to environmental agents. The question of how much cancer comes from a simple stochastic event (bad luck) and how much cancer has identifiable specific causes has been a matter of much discussion and speculation. But no one had attempted to find a mathematical solution to the problem from a purely biological reference point before Vogelstein and Tomasetti.

Previous attempts to quantify the combination of genetic, lifestyle, and chemical factors compared cancer rates in different countries with the presumption that the country with the lowest rates defined what was "unavoidable" and that the higher rates in other countries were attributable to some exogenous or genetic cause. Doll and Peto had estimated in 1981 that 75 to 80 percent of cancer could have been avoidable, which would leave about only one-quarter of cases caused by bad luck and genetics.[29] Following that math, if the National Cancer Institute concluded that genetics contributes only 5 to 10 percent, this would leave less than 20 percent caused by bad luck, according to Doll and Peto.[30]

It should be noted that both methods of approaching the "bad luck" problem have their limitations. In any event, we can conclude that both preventable and nonpreventable causes are substantial contributors to cancer. As toxicologists we rightly focus on the preventable causes, especially chemicals and cigarettes. Nutritional epidemiologists and scientists examine foods, nutrients, and micronutrients. Infectious disease researchers and radiation biologists study their areas. We each have our part in trying to prevent cancer. According to Ernst Wynder, "What we eat and drink, what we smoke, our physical inactivity, our sexual practices, our use of illicit drugs and even our excessive exposure to sunlight put us at risk for often incurable illnesses."[31]

How Do We Use Toxicology?

One ought to be afraid of nothing other
Than things possessed of power to do us harm
But things innocuous need not be feared.
—Dante, *Inferno*

This group of chapters begins by showing how the results of toxicological studies can be used to inform the environmental and occupational regulations that protect workers and the public. This has worked well, and regulation has been particularly successful at protecting workers from unacceptable exposure doses; as we recall, Paracelsus observed centuries ago that the dose makes the poison. Regulatory procedures involved in cleaning up contaminated sites have provided another useful product of toxicology, although in some instances this has occurred without commonsense solutions to problems. The history of the development of regulatory standards and cleanup levels for chemicals also raises questions about the uses and misuses of animal bioassays to determine whether and at what dose chemicals become a threat to humans.

The last two chapters of this group describe topics that do not involve most toxicologists. The first is the use of toxicology in the courtroom by expert witnesses, to determine whether a person's illness or death was caused by a chemical exposure. The second is the toxicology of war, where toxic substances intentionally have been used to harm military personnel and even civilians. This subject is usually not included in textbooks or taught in toxicology courses, but toxicologists were involved in the development of these chemicals and also in the development of antidotes to counter their effects.

15 | Protecting Workers from Chemical Diseases

Workers have suffered the greatest exposures to chemicals and, consequently, chemical diseases. As mentioned earlier in this book, mining's serious occupational hazards were described by the sixteenth-century physicians Paracelsus and Agricola. Agricola described the first efforts to diminish industrial hazards through the proper construction of mines and ventilation. Development of these systems was aided by the seventeenth-century physician Bernardo Ramazzini, who studied occupational exposures in various professions and discussed ways of mitigating the hazards. Percival Pott and Henry Butlin described scrotal cancer among chimney sweeps and noted the protective effects of personal hygiene. Centuries later, what has become a standardized system of workers' protection is called "industrial hygiene" and provides for the identification and prevention of causes of injury or disease.

Silicosis was the most prevalent occupational disease in most mining operations. One of the first major improvements for occupational health was the invention of wet drilling by J. George Leyner, of Denver, Colorado. The improved drill, patented in 1897, forced water into the drill hole, creating a harmless mud out of the otherwise dusty and airborne drill shavings.[1] Other industries also saw advances in occupational health. In the United Kingdom, the Cotton Cloth Factories Act of 1889 set a limit for carbon dioxide levels of nine volumes for every thousand volumes of air. This required factories to be ventilated by fans and prompted the development of sampling devices for carbon dioxide. [2]

In the United States at the beginning of the twentieth century, Dr. Alice Hamilton pioneered occupational medicine, or "industrial medicine," as it was called at that time. Hamilton is not a well-known figure, but she was as important to occupational toxicology as Rachel Carson was to ecological toxicology in the 1960s. She developed occupational epidemiology and industrial hygiene, drawing from the discoveries being made in the emerging laboratory science of toxicology. Hamilton was the first woman to be on the faculty of Harvard University; she was appointed in 1919 to assistant professor of industrial medicine. She also served two terms on the Health Committee of the League of Nations and was a consultant to the U.S. Division of Labor Standards.[3]

After some work in 1902 related to bacterial outbreaks such as typhoid fever, Hamilton realized that little had been written or was understood in the United States about occupational illnesses as a public health problem. There was some discussion of this subject in Europe, but both U.S. industry and the medical profession had turned a blind eye to the diseases caused by the workplace. The American Medical Association had never held a meeting devoted to industrial medicine. There was a mistaken acceptance of the idea that the conditions described in Europe did not occur in America.[4]

Hamilton doubted that there was a lack of occupational dangers in America, and her opinion was confirmed in 1908 when John Andrews described to her a condition called "fossy jaw," which had been well known in Europe for over fifty years. This debilitating disease was caused by exposure to breathing white or yellow phosphorus, which was used to make match tips. The airborne phosphorus would penetrate a defective tooth and destroy the jaw, leading to an abscess. Cases of workers exhibiting the toxic effects of phosphorus had been found in the United States and Europe in about 1845, but whereas Europe had actively discussed and taken preventative measures, the American medical establishment was silent.[5]

Lead poisoning was the next illness examined by Hamilton when she began her work for the Illinois Occupational Disease Commission in 1910. She discovered cases of poisoning from the manufacture of white lead pigment produced from metallic lead. This white lead was mixed with turpentine and linseed oil to make paint. Most of the exposed workers were new immigrants to the United States who were given the dirtiest and most hazardous jobs. She described the case of a thirty-six-year-old Hungarian who spent six years grinding lead paint, during which time he had three episodes of colic with vomiting and headache. When she saw him in the hospital, he was a skeleton of a man: he looked twice his age and his muscles had atrophied. Another worker of Polish origin had only had three weeks of exposure, in a very dusty job in a white lead plant, when he was hospitalized with lead colic, paralysis, and tremors in his wrists.[6]

Hamilton's method of changing the work environment involved educating the workplace owners. In some cases, she was successful, and the employers made changes that greatly reduced exposures when they learned about the diseases being caused by lead. Other times it was not so easy. For many decades, it was common to add lead to porcelain enamel glazes as a bonding or pigmenting agent. Hamilton discovered cases of lead poisoning in workers who were applying the lead to heated iron, in the manufacture of bathtubs. In one case described by Hamilton, a bathtub enameller worked for eighteen months and then passed out at the furnace, remaining in a coma for four days. After regaining consciousness, he was delirious and partially paralyzed in both his arms and legs.[7]

During World War I, Hamilton examined the munitions industry. In order to produce nitrocellulose and other nitrated explosives, large quantities of nitric acid were required. Nitrous fumes caused irritation of the lungs, leading workers to drown in their own fluids. In the list of 2,432 cases of industrial poisoning that Hamilton published in 1917, nitrous fumes accounted for 1,389 cases and for twenty-eight of the fifty-three deaths. TNT came in second, with 660 cases and thirteen deaths. In her autobiography first published in 1943, *Exploring the Dangerous Trades*, Hamilton wrote:

"It is hard to believe that this rich and safe country should refuse to give its munition workers the sort of protection which France and England, fighting for their lives, provided as a matter of course. But it was impossible to overcome the arrogance of the manufacturers, the indifference of the military, and the contempt of the trade-unions for non-union labor."[8]

During World War I, there was high demand for mercury fulminate, a component of detonators. In most cases, mercury is found in the ground as mercury sulfide, also called red cinnabar, which is relatively harmless to miners. However, some mines contain pure mercury, or droplets of quicksilver, and in the hot conditions of the mines, the vapors caused mercury poisoning. Hamilton found that the drilling of the rock to place the dynamite and the subsequent explosion produced a fine dust full of tiny droplets of mercury. The poisoning was not hard to detect—it was the same as "mad hatter's" disease, which had been described by Ramazzini and others. Hamilton not only found mercury poisoning in miners but also in felt hatters and makers of thermometers, dry batteries, and dental fillings. In the case of mercury, some of the exposures could be decreased by personal hygiene and clothes washing.[9]

After World War I, Hamilton investigated carbon monoxide poisoning among miners who were exposed when coal mines were blasted. The detonations were performed after the end of the shift, so that the poisonous gas could disperse with time; however, too often the gas did not disperse sufficiently, and consequently many miners asphyxiated. Carbon monoxide is toxic because it combines with hemoglobin, preventing the transport of oxygen from the lung to the tissues in the body. Carbon monoxide binds much more strongly to hemoglobin than oxygen and thus is very efficient at suffocating an exposed person. Another industry with carbon monoxide exposure that Hamilton investigated was in the making of steel using the production of coke for fuel from coal.[10]

The biggest source of carbon monoxide exposures for most workers would come from the internal combustion engine. As an example of how this problem persists, Pennsylvania Station in New York was rebuilt in the 1960s after the demolition of the classic original glass structure. The redesigned station is underground, underneath Madison Square Garden. It was supposed to accommodate only electrified trains. All diesel engines were to be disconnected and reconnected in Washington, New Haven, and Croton Harmon so that only electric trains would travel between these stations and New York. However, changing the locomotives proved impractical, and diesel engines wound up in Pennsylvania Station. As a consequence, ventilation systems had to be retrofitted to decrease the diesel exhaust, which in some cases did not work well. High levels of carbon monoxide persisted, especially in some of the tunnels.

• • •

Besides reducing exposures in the workplace, another aspect of occupational toxicology involves systematically keeping track of the potential effects of chemicals on workers. In 1933, J. C. Bridge proposed that one of the most important—if not the most important—means of protecting workers was medical supervision, which was still a comparatively new idea. Under the United Kingdom's Factory and Workshop Act, certain diseases were notifiable by medical practitioners to the chief inspector of factories if contracted in a factory or workshop.

Lead poisoning was still the biggest culprit; during the years 1900 through 1931 there were 1,736 deaths attributed to chronic nephritis from lead exposures. The largest number of lead poisoning cases in any one industry was sixty-four, in 1931, out of a total of 168, occurring among painters of buildings using lead paint. Dr. Bridge also noted that the risk of silicosis had been improved through ventilation and cleanliness. Dermatitis was also a major occupational concern and was not considered to be controllable solely through the use of gloves. Periodic examinations of workers exposed to silica and asbestos were also performed.[11] These examinations also found cases of tuberculosis among workers.

Another more recent example of medical surveillance involves vinyl chloride workers and the precautions taken following the identification of the risk of angiosarcoma of the liver, a deadly cancer. To ensure that the stringent regulation of the allowable air level is not breached, workers are given regular blood tests to detect any hint of liver toxicity. Vinyl chloride will show liver toxicity; this is reversible if exposure is stopped by removing workers from the workplace. They are not allowed to return to the workplace until their biomarkers of liver toxicity are normal and the conditions that produced the increased levels of vinyl chloride in the air are corrected.

In the United States, occupational toxicology became organized when the National Conference of Governmental Industrial Hygienists (NCGIH) convened on June 27, 1938, in Washington, DC. Representatives to the conference included seventy-six members, representing twenty-four states, three cities, one university, the U.S. Public Health Service, the U.S. Bureau of Mines, and the Tennessee Valley Authority. This meeting was the culmination of concerted efforts by John J. Bloomfield and Royd S. Sayers. Bloomfield had performed some of the leading research on silicosis and black lung disease of coal miners and then traveled to South America, where he established industrial hygiene programs. Sayers was chief of the Division of Industrial Hygiene at the National Institutes of Health from 1933 to 1940 and director of the U.S. Bureau of Mines from 1940 to 1947.

In 1946 NCGIH changed its name to the American Conference of Governmental Industrial Hygienists (ACGIH). Undoubtedly the best known of ACGIH's activities, the Threshold Limit Values for Chemical Substances Committee was established in 1941. This group was charged with investigating, recommending, and annually reviewing air limits for chemical substances. It became a standing committee in 1944. Two years later, the organization adopted its first list of 148 exposure limits, referred to as "maximum allowable concentrations." The term "threshold limit values" (TLVs) was introduced in 1956. The first edition of *Documentation of the TLVs* was published in 1962, and it is updated and published every year.[12]

Finally, in response to dangerous working conditions in other industries, the U.S. government acted. The bipartisan Williams-Steiger Occupational Safety and Health Act of 1970 was signed into law by President Richard M. Nixon. This law led to the establishment of the Occupational Safety and Health Administration (OSHA), the National Institute of Occupational Safety and Health (NIOSH), and the independent Occupational Safety and Health Review Commission. George Guenther became the first assistant secretary of labor for occupational safety and health under President Richard Nixon. He oversaw the adoption of OSHA's first standard on asbestos in 1972. Standards for thirteen other carcinogens were adopted in 1974.[13]

OSHA set permissible exposure limits (PELs), which, if not exceeded in the workplace, minimize the harm to workers. These limits are enforceable by the government, and if they are exceeded in the workplace and can't be reduced, workers are supposed to wear personal protective equipment such as masks, gloves, and special clothing. We have all seen pictures of workers in white Tyvek suits involved in cleaning up a hazardous spill. However, the primary goal of OSHA PELs, as with the ACGIH limits, is to provide for solutions that decrease the air levels in the workplace so that protective clothing is not necessary.

OSHA standards are not designed to protect workers from any disease completely but rather minimize the possibility of disease. OSHA regulations also have to take into account financial issues, to prevent the regulations from becoming so stringent that they are cost prohibitive. Many jobs involve risks, and the wages of workers are supposed to include compensation for risks. It is a matter of debate whether certain jobs adequately pay for risks, but in general, in the post–World War II era, we have gradually accepted fewer and fewer risks as a condition of employment.

OSHA currently has PELs for nearly five hundred hazardous chemicals. These were established in 1971, when the agency was created, and are based on research conducted primarily in the 1950s and early 1960s. Since then, much new information has become available, indicating that, in most cases, these early exposure limits are outdated and do not adequately protect workers. But, unlike the ACGIH, OSHA faces legal challenges when setting new PELs, and in most cases these have been impossible to change.

The protection of workers has been a difficult task for the federal government, given the complexity in setting these standards. Fortunately, industries usually try to comply with the ACGIH recommendations; otherwise, they get challenged by the workers and even sued by worker's lawyers. In the United States, the ACGIH provides the most extensive guidance for the protection of workers from the toxic effects of chemicals. Although its limits on exposure have no legal authority, they are widely used by industry. Also, because they do not represent government regulations, they can be developed and changed without the legal battles that have plagued OSHA.

One of the first listed of the 148 exposure limits set by the ACGIH was for benzene. In 1946, the TLV was 100 ppm; this was lowered several times over the years, and in 1997 it was finally decreased to 0.5 ppm. Originally, the TLV was based on the most obvious and highest-dose toxic effect of benzene, aplastic anemia, whereby the bone marrow stops producing blood cells. As information about benzene's leukemogenic properties became known, the final determination was made that represented a two-hundred-fold decrease in the allowable air level. Today's list of TLVs includes over seven hundred chemical substances and physical agents. ACGIH has additionally established levels of fifty exposure biomarkers, called "Biological Exposure Indices" (BEIs), for selected chemicals.

• • •

Protecting workers has not been straightforward in some cases. A case in point has been the air standard for beryllium. Given its mechanism of causing disease, some people have a much greater sensitivity for developing berylliosis, also known as chronic beryllium disease (CBD), than others. In a dramatic announcement in April 2000, Bill Richardson, secretary of the U.S. Department of Energy (DOE), acknowledged that his agency had helped the beryllium industry defeat a 1975 attempt by OSHA to reduce workers' exposure to beryllium. Richardson also stated that "priority one was production of our nuclear weapons" and that the "last priority was the safety and health of the workers that build these weapons."[14]

Beryllium is a metallic element with the atomic number of 4. Although rare, its gemstone forms include aquamarine and emerald. Its less precious ores are mined and processed into various forms, including the pure metal, alloys, and ceramics. Purified elemental beryllium appears as a fine gray powder with a metallic luster. Given its low atomic number, beryllium has a small ionic radius, and its chemical properties stem from its high-density charge. It is lightweight, has a stiffness six times that of steel, has low density, and its high melting point is 1,285°C, all of which make it a strategically important metal.[15]

The most important use of beryllium was in the manufacture of nuclear weapons. Beryllium complements plutonium in the development of a nuclear fission reaction by being a neutron donor. The plutonium bomb known as "Fat Man" was one of the two types of nuclear bombs developed, and it was the one dropped on Nagasaki.[16] This bomb contains a beryllium core, surrounded by plutonium and then uranium, which in turn is surrounded by an explosive lens weighing about two and a half tons. The detonation is an inward implosion, mixing the plutonium and beryllium under immense pressure. This creates a cocktail of alpha particles and neutrons from the beryllium that causes fission of the surrounding uranium.[17]

To use the metal, it had to be machined, creating dust in the working environment. The first cases of chemical pneumonitis associated with beryllium exposure in the United States were reported in 1943 by H. S. van Ordstrand, and the exposures occurred while workers were extracting beryllium oxide from beryllium ore. By 1945 this group of researchers had seen 170 cases of acute beryllium poisoning. The symptom complexes included dermatitis, chronic skin ulcers, as well as pneumonitis; five of these workers had died.[18]

In 1950, the Atomic Energy Commission assembled a committee of experts to recommend safe levels of beryllium in the air. Other than rare cases resulting from accidental exposures, the acute form of beryllium disease was no longer a problem in the United States. Inhaling high concentrations of beryllium led to acute chemical pneumonitis, producing difficulty breathing, cough, and chest pain. This acute process occasionally results in death, but when the patient survived the acute illness, recovery was thought to be complete.[19]

The occupational air standard for beryllium, 2 micrograms per cubic meter of air averaged over an eight-hour work day, was first proposed in 1951. This air level was not based on the toxicity of beryllium but on that of other metals, such as arsenic, lead, and mercury. This relatively low level was set to reflect beryllium's lower atomic weight and concern about its greater toxicity. The allowable level was considered to be quite safe, and fifty years later this standard was the one OSHA enforced in the workplace.[20] With advances in industrial hygiene, the acute form of beryllium disease was virtually eliminated in the United States using the standard of 2 $\mu g/m^3$. There were fifty-three acute cases reported in 1947, twenty-eight in 1948, and only one in 1949.[21]

However, another disease caused by beryllium was discovered: chronic beryllium disease (CBD). This condition resulted from prolonged or multiple air exposures to lower levels of beryllium. CBD resembles tuberculosis in that some people develop immunologic sensitization to beryllium over a period of months or years, producing granulomas in the lungs that eventually limit breathing capacity. At first the OSHA standard was believed to protect workers from this disease as well. Cases of CBD were blamed on exceedances of the standard through accidents in the workplace. Another problem is that CBD, like cancer, has a long latency; cases do not occur until decades after initial exposure.

In 2004 NIOSH estimated that up to 134,000 workers in the United States had been exposed to beryllium and were at risk of the chronic form of the disease. The OSHA PEL for beryllium in air, 2.0 micrograms per cubic meter, came to be recognized as failing to protect some exposed workers from sensitization and eventual CBD. A blood test was eventually developed to identify workers who became sensitized to beryllium and thus needed to be removed from the workplace before developing this disease. A growing body of evidence indicated that a genetic susceptibility of the exposed workers contributed to beryllium disease outcomes.[22] OSHA issued a proposed rule in 2015 to reduce the PEL for beryllium in air tenfold, to 0.2 micrograms per cubic meter, followed by a months-long public comment period and several days of public hearings. This rule was finalized on May 20, 2017.[23]

Protecting workers from diseases of chemical exposures has come a long way in the last hundred years and has probably been toxicology's most complete achievement. By that I mean that the identification of hazards has reduced harmful exposures that could lead to health effects such as cancer. OSHA regulations and ACGIH exposure limits primarily are derived from studying workers rather than relying on animal testing or other indirect measures. Of all the venues in which toxicology operates, occupational toxicology has probably been the most successful in preventing the gravest diseases caused by chemical exposures.

16

The Importance of Having a Good Name

I t is a commonly held misconception that manmade chemicals are major risks for cancer. As we saw with Doll and Peto's
estimates, according to epidemiologists, this is not the case.
Of the thousands of chemicals tested for cancer in animals and
studied in humans by epidemiologists, as of 2019 only seventy-
two chemicals and other agents have been found conclusively
to cause cancer in humans by the World Health Organization's
International Agency for Research on Cancer (IARC). This
includes thirty chemicals that are synthetic or produced by
industrial processes, six naturally occurring agents, nineteen
drugs, ten mixtures (for example, soot), and thirteen occupations
that are considered "human carcinogens." There are more than
four times as many agents found to cause cancer in rodents but
not classified as human carcinogens; most of these are industrial
synthetic chemicals.[1]

The small percentage of human cancer caused by chemicals
does not mean that we do not need to be vigilant or protect ourselves against exposures to them. It does mean, however, that
when a toxicologist is faced with determining whether a person's
cancer could have come from exposures to specific chemicals,

that evaluation must be conducted in the context of understanding all other possible sources of cancer, including smoking, obesity, infection, and bad luck. According to the IARC, human carcinogens include seven lifestyle causes, including tobacco and alcohol; fifteen types or sources of radiation; six viruses; two parasites; and one type of bacteria.[2]

A major problem with the perception regarding chemicals causing cancer is that animal testing has identified hundreds of chemicals that cause cancer in rodents but not in humans. Liver cancer has been the most common cancer found in test animals caused by industrial chemicals, and this finding is the most common reason for classifying a chemical as a carcinogen.[3] If one assumed that regulations based on finding liver carcinogens would have a protective effect, one would expect that liver cancer rates in humans should be dropping dramatically. But the opposite has happened. Despite the reduction in industrial chemical pollution using regulations based on liver tumor findings in rodents, liver cancers have been increasing over the past two decades. If this increase is not because of higher exposures to chemicals, there must be another explanation. As described in a previous chapter, the major risk factors in the United States for liver cancer are excessive alcohol use and hepatitis infections, both of which have increased.

• • •

Over the last century, benzene, asbestos, and cigarettes were found to be notable causes of cancer and other diseases. Understanding their means of causing cancer was pivotal in the history of toxicology. Cigarette smoking was and remains the most important cause of lung cancer and other preventable diseases such as emphysema and heart disease. Asbestos was and is considered to be the most important cause of occupational lung disease, causing cancer and pneumoconiosis. Benzene was a widely used industrial solvent and synthetic chemical precursor, causing bone marrow toxicity and acute myelogenous leukemia; however, its use and contamination in products has been almost eliminated.

But amazingly, the public's concern about these chemicals was dwarfed by an environmental threat: polychlorinated biphenyls (PCBs). PCBs were invented by German chemists and produced in the United States by Monsanto. They were used as the dielectric insulation between metal wires carrying electricity because they were much more fire resistant than the mineral oils that had been used previously. They were used as insulation for electrical wires and in transformers and capacitors. For example, a mixture of PCBs called "cablewax" insulated electric cables in ships and submarines during World War II. After the war, PCBs insulated the components of large commercial transformers—those changing the transmission-line power of 10,000 volts to 220 or 110 volts. PCBs worked well and lowered the number of fires that had plagued the previous generation of high-voltage electrical equipment, which were insulated using mineral oils.

Why were PCBs such a media sensation? Besides having an easy-to-remember name, a couple of well-published poisonings occurred from PCB products produced in Japan: the Yusho and the Yu Cheng episodes. In 1968, rice oil in Yusho, Japan, was being heat treated in a heat exchange system using PCBs in the heating coils. Cracks developed in the coils, and the tainted rice oil was then used for cooking. At least one thousand people became ill after consuming that oil over a several-month period. Various health effects were reported, including skin, neurologic, liver, and eye toxicity, especially in the babies born to exposed mothers. Another case of similar rice oil poisoning occurred in Yu Cheng, Taiwan, in 1979. Chemical analyses and toxicological investigations revealed that the main culprits were PCB breakdown products called polychlorinated dibenzofurans (PCDFs), caused by the prolonged heating of the PCBs in the heat exchange system. The PCDFs were much more toxic than the PCBs.[4]

One characteristic that made PCBs environmentally hazardous also made them useful for industrial applications: they were highly resistant to degradation. This also meant that once they were in the soil, in the sediment, in the water, or in the body of an animal, they were there to stay almost permanently. Once in a body of water, for example, they were consumed by various invertebrates, which were consumed by small fish, who were then eaten by bigger fish. Another problem for PCBs was that, like DDT and the other chlorinated pesticides, PCBs would biomagnify, becoming more concentrated as they moved up the food chain. In 1971 fish in Alabama were found to contain levels of PCBs of up to 227 parts per million;

the Food and Drug Administration's limit in fish at that time was only five parts per million.[5] These PCBs came from a Monsanto plant nearby, where forty-five tons had been discharged into the water in 1969 alone.[6] And in 1970 it was found that chickens in the United States were contaminated with PCBs from a processing system for fish meal used for feed.[7] Charles C. Edwards, then head of the Food and Drug Administration, sought to calm fears. He said that there was public confusion caused by "a few alarmists seeking headlines and abetted by unbalanced reporting."

In 1975 PCBs were found to cause liver cancer in mice and rats by Renate Kimbrough, at the Centers for Disease Control.[8] They also had other remarkable biological properties. One was the ability to induce the production of cytochrome P450 in the liver, and, in fact, as we saw earlier, they were so good at doing this that Bruce Ames incorporated the metabolic activation system of liver enzymes from rats injected with PCBs in his standard mutagenicity test protocol.

PCBs had much more help becoming famous. They were put in large electrical transformers because they were not supposed to burn, but a PCB-containing transformer caught on fire on February 5, 1981, in a large federal office building in Binghamton, New York. Governor Carey in a publicity stunt said, "I offer here and now to walk into Binghamton or any part of that building and swallow an entire glass of PCBs." He also stated, "If I had a couple willing hands and a few vacuum cleaners, I'd clean that building myself."

The entire building was contaminated by PCBs and by some of the same extremely toxic chlorinated furans that were found in the Yusho and Yu Chen rice oils; chlorinated dioxins had also been formed. This episode put the public health concerns about dioxins on the map. The workers who decontaminated the Binghamton office building had to wear white Tyvex suits and masks and conform to strict safety procedures. The cleanup took thirteen years, eight months, and six days, and $53 million later, state officials finally declared the Binghamton office building safe to reoccupy. The first of 630 tenants arrived on October 11, 1994: two judges and six staff members assigned to the State Courts of Claims.[9]

PCBs were also found in Hudson River striped bass and in salmon from Lake Ontario.[10] Commercial fishing on the Hudson was closed and remains so to this day. PCBs came from the General Electric plant in Hudson Falls, where PCB-containing electrical equipment was manufactured. More than any other contaminant, PCBs have restricted recreational fishing activities in the United States: fish advisories warn people not to eat the fish or to do so only in small quantities. Most PCBs are barely soluble in water, another reason for their persistence in river sediments. There were restrictions placed on the size of striped bass that sport fishermen were allowed to keep because of PCB levels. The commercial fishing ban was, unexpectedly, a tremendous boon for the fish population. When the commercial fishing ban started, the striped bass population had dwindled to record-low numbers; after the ban their population surged. Apparently, the fish suffered no toxic effects from the PCBs.

PCBs were only one of the favorites of the press that my boss Ernst Wynder would call "the carcinogen of the week." He wondered why the press didn't give equal time, for example, to cigarette smoking. Where, he asked, was the evidence that PCBs actually caused human disease? At that time, no one had proven that PCBs caused any form of cancer or other disease in occupational exposures except for the dermatological condition chloracne. This was a particularly severe and scar-forming type of acne, which had earlier been found in only a small number of PCB-exposed workers. Other effects found in the Yusho and Yu Cheng episodes required exposures to the much more toxic heat breakdown products of PCBs, not the PCBs themselves.

But because they had been associated with cancer in rodents and widespread contamination, huge amounts of government funding were being spent to research PCBs. As of 1996, according to the National Library of Medicine's PubMed database, there were almost five thousand publications involving PCBs. Ernst was furious that these monies were not being spent on cigarette smoking prevention, since cigarettes had been proven to cause at least one-third of all human cancer in the United States. It seemed to be further evidence that either the tobacco companies were having their way or that the government cared more about "hot topics" than about public health. Wynder believed that the government was being duped and that lives were at stake. As of 2018, the number of PCB publications has risen to twenty thousand, compared to a bit over forty thousand for cigarettes.

Wynder was an advocate of Gary Williams's studies at the American Health Foundation because they provided some counterbalance to what he believed was an overzealous emphasis on hypothetical human carcinogens. Studies of industrial chemicals causing cancer or other diseases only at very high exposures in experimental animals were overshadowing his research of tobacco chemicals, which were causing an enormous cancer burden in humans. Williams was a pioneer in showing that environmental exposures to most animal carcinogens would have little or no impact on human cancer. Wynder contended that human exposure to industrial chemicals in the general population was small compared to the direct inhalation of cancer-causing chemicals in cigarette smoke. In his view, the media greatly exaggerated the health effects of industrial chemical exposures, privileging sensationalism over good science. In doing so, they repeatedly made us aware of health issues of lesser importance while ignoring the more commonplace yet more important causes of disease, such as cigarette smoking, obesity, and infection.

• • •

Evidence from studying certain human exposures in occupations supported by animal studies suggested decades ago that we were on the verge of a major breakthrough in preventing cancer. Although adequate studies of many chemicals in humans were not possible, studies in rodents led to the identification of hundreds of "cancer-causing chemicals." During the 1970s and 1980s, the media reported about polluted places: Love Canal, Times Beach, the Hudson River, Woburn, and so on. The concerns about the toxicity of the chemicals involved were based largely on animal studies. Reading the newspapers during this time, it seemed as though a new major source of human cancer was being discovered every week. If only we could get the Alar out of apples, saccharin out of soft drinks, PCBs out of fish, and dioxin out of everywhere, pundits seemed to suggest that we wouldn't get cancer at anywhere near the rates that we had in the past. This was well meaning, but it was not clear that the major cause of cancer in humans was low-level environmental exposure to chemicals. The public was more scared about unfamiliar things that might hurt them than the familiar things, like cigarette smoking, that were certainly killing them.

More attention should have been given by the media to the risk that cigarette smoking contributed to other diseases besides cancer, as well. The Centers for Disease Control and Prevention (CDC) estimate that 24 million Americans have emphysema. In the United States, emphysema is the fourth-leading cause of death, and cigarette smoking is the major cause of emphysema. This was also true when Wynder was promoting smoking cessation.

Meanwhile, according to the American Lung Association, "Every year tobacco kills more Americans than did World War II—more than AIDS, cocaine, heroin, alcohol, vehicular accidents, homicide and suicide combined. Approximately 443,000 people die prematurely from smoking or exposure to secondhand smoke each year."[11] Wynder's fury was justified. The challenge remains how to reconcile the public's perception of risk with the scientifically verified risk from different agents. Alone with me, he tended to flail around angrily and with passion, but he was much more cool and effective when facing an audience. Wynder could be very charming, and his European accent added to his sense of authority.

Fortunately for us, Wynder's message has now become more the norm. Today, the public and the media have focused on more important issues in public health: smoking, obesity, infection, and global warming. Not that we need not be aware of the potential of industrial chemicals to do harm, but a more educated perspective regarding risk seems to be the wave of the future. Were Wynder alive today, he would certainly still be concerned about the damage that people do to themselves from cigarettes.

17 | Can We Accurately Regulate Chemicals?

eginning in 1969, the International Agency for Research on Cancer (IARC), which is part of the World Health Organization (WHO), established the first program to evaluate the cancer risk that chemicals, other agents, and certain occupations pose to humans. The IARC is funded by agencies of participating world governments, such as the U.S. National Cancer Institute. The IARC program entails the preparation and publication of monographs evaluating the carcinogenic risk of individual chemicals, defined chemical mixtures, other agents such as radiation and viruses, and certain occupations (such as working as a painter or rubber worker). These monographs are now known as the *IARC Monographs on the Evaluation of Carcinogenic Risks to Humans*. The overall evaluation in the monograph provides a group number indicating how certain the working group believes the evidence of carcinogenicity is in either humans or animals. A Group 1 designation means that the agent or occupation is a known human carcinogen.

The monographs are prepared by working groups made up of primarily academic and government scientists from countries around the world. The IARC is not a regulatory body, but it provides the information governments can use to regulate carcinogens. Many countries adopt the IARC cancer classification wholesale; others have their own classification procedures. In the United States, the Environmental Protection Agency (EPA) and National Toxicology Program (NTP) classify chemicals for regulatory purposes. The NTP published the first classification of chemicals in its *Review of Carcinogens* in 1980, and the EPA did so somewhat later in the 1980s.

From its inception until 1990, the cancer classification system used across the board by the IARC, EPA, and NTP considered human evidence for cancer based on a scientific interpretation of the results of previous studies. If a chemical was shown to cause cancer in humans in epidemiology studies, it was classified as a human carcinogen. But how does one show that a substance causes cancer? Take the example of benzene and leukemia. It is much more difficult to link leukemia to benzene exposure than it is to link mesothelioma to asbestos or vinyl chloride to angiosarcoma of the liver. This is both because leukemia is not a rare disease and because only a relatively small percentage of workers were afflicted with this cancer. So toxicologists were faced with the question: how does one prove that benzene causes leukemia, specifically, acute myelogenous leukemia (AML)?

Early reports of leukemia in workers using benzene raised suspicions, but they could likely have been chance occurrences. Investigators in the USSR in 1963 described sixteen cases of leukemia in workers occupationally exposed to benzene. Around the same time, Drs. Enrico C. Vigliani and Giulio Saita, from the Clinica del Lavoro of the University of Milan, Italy, reported in 1964 on the possible association between exposure to benzene and leukemia. They found forty-seven cases of blood abnormalities, of which six were leukemia patients who reported exposures to benzene. At the Institute of Occupational Health in Pavia, forty-one cases of blood abnormalities were seen between 1961 and 1963 (including five leukemias) among workers exposed to glues containing benzene used in shoemaking.[1] But since Vigliani and other investigators did not compare their cases to matched control groups, as Wynder and Doll had done in their studies on cigarette smoking and lung cancer, one could not be certain that these findings were not a coincidental relationship or possibly attributable to other causes.[2]

As with the study of cigarette smoking and lung cancer, what was needed were epidemiology studies that compared the number of cancers among workers exposed to benzene to those occurring in unexposed people. Richard Doll's collaborator Austin Bradford Hill had established a methodology to approach the evaluation of causality based on epidemiological and other types of studies. In his president's address to the Royal Society of Medicine, Hill explored the question of when a statistical association could prove causation for studies of cigarette smoking and lung cancer. To make this distinction, Hill asked, "What aspects of that association should we especially consider before deciding that the most likely interpretation of it is causation?" Although Hill developed his method to prove that cigarette smoking causes lung cancer, it can be used to address any argument for chemical causation of disease, for example, benzene and leukemia.[3]

The first issue Hill considered was the *strength* of the association. As an example, he pointed to the enormous increase in the incidence of scrotal cancer in chimney sweeps described by Percival Pott. Hill reported that the mortality of chimney sweeps from scrotal cancer was two hundred times greater than workers not exposed. He also cited the tenfold increase of lung cancer in cigarette smokers. He was not, however, summarily impressed with the twofold increase in coronary thrombosis in cigarette smokers, because all of the other possible factors that could cause this disease made it difficult to control properly for these in an epidemiology study.

Next on Hill's list of factors to be considered was *consistency* of the observed association. In other words, has the association been observed repeatedly by different persons in different places, circumstances, and times? Consistency among studies would rule out statistical chance as the cause of the positive association, and consistency among studies with different features would argue that there was no constant error or fallacy among the different studies.[4]

Specificity was the third characteristic Hill considered, and by this he meant that the association was limited to specific workers and a particular type of disease. If a chemical causes cancer, it usually causes only one or a few different types of cancer. Cigarette smoking may be an atypical case because of all the different carcinogens involved.

Hill's fourth characteristic was *temporality*. The chemical exposure must have occurred before the development of the disease. For occupational chemical exposures, temporality is usually obvious from the work history of the affected individual, but for environmental exposures this may be tricky. The fifth feature is what Hill called *biologic gradient*, which would be later called dose response. In the case of cigarette smoking, for example, Ernst Wynder's study found that the more cigarettes smoked, the stronger the association.[5]

Plausibility was a feature that Hill said could not be demanded because it depends upon the knowledge of the day. Although Hill did not describe in detail what he meant by this factor it has been interpreted to include an understanding of the mechanism by which the cancer is produced based on experimental studies. A feature similar to plausibility was *coherence*, which meant that the cause-and-effect relationship should not conflict with known facts about the natural history of disease. In the case of cigarette smoking, the temporal rise of lung cancer coincided with increases in cigarette smoking, and the sex difference in lung cancer and smoking contributed to coherence because at that time most men smoked whereas most women did not. Wynder's demonstration that tobacco tar produced skin cancer in experimental animals and the identification of benzo[a]pyrene in cigarettes would have lent coherence to the association; today it might be argued that this contributes to biological plausibility instead of coherence.

Experiment rarely contributes, but this feature is possible if an exposure is prevented and the frequency of associated events is affected. In some circumstances Hill believed it would be fair to judge by *analogy*. He cited the birth defects caused by thalidomide or rubella as lowering the bar to consider exposure to another drug or viral infection during pregnancy to cause birth defects. Earlier, the possible causes of birth defects were unknown, but now our understanding that thalidomide and rubella can cause birth defects gives credibility to arguing that other drugs or viral infections, not just thalidomide or rubella, might also.[6]

According to Hill: "Here then are nine different viewpoints from all of which we should study association before we cry causation." However, he cautioned against using these as hard-and-fast rules of evidence that must be obeyed before determining cause and effect.[7] For example, the causal link between vinyl chloride and angiosarcoma of the liver was quickly accepted following the publication of just two small studies because the disease was so rare and the exposures in the workplaces were consistent and restricted to primarily vinyl chloride. For benzene and AML, by contrast, it would take several studies to establish a cause-and-effect relationship to meet the requirements of the analysis required by Hill.

• • •

Cancer classification systems also evaluate the results of animal tests. However, though the evaluation is seemingly straightforward, it lacks in scientific judgment regarding whether the animal testing results are reliable. If a chemical caused a statistically significant increase of tumors in animals in two independent and well-conducted studies, it is found by regulatory agencies to be "carcinogenic to animals" no matter how many studies didn't show this result.

The difference between the evaluation of human and animal studies is quite dramatic because of the robust set of criteria needed for acceptance of human studies. The Bradford-Hill type of analysis in practice at the IARC excludes many epidemiological studies from consideration because of the quality of the study or the possibility of what are called confounding factors and perceived bias in the data. In other words, there is more judgment allowed in the evaluation of the human studies, whereas for the animal studies, it becomes mostly a counting exercise: two strikes and you're a carcinogen. Based on such animal study results, chemicals are given the designation of "possible, probable, or likely to be" carcinogenic to humans, and these descriptors may not be appropriate in many cases.

After this designation is formulated, an amazing thing happens in the use of any specific cancer classification in the United States by the EPA and NTP. It doesn't matter whether a chemical has been proven to be a human carcinogen or is only an animal carcinogen; they are all treated the same in terms of government regulations for chemicals in the environment. Governmental agencies believe that it's reasonable and prudent to expect that an agent that causes cancer in animals may also cause cancer in humans under some conditions of exposure. However, one could also make the argument that Group 1 known human carcinogens should have greater regulatory oversight.

For example, when PCBs were found to cause liver cancer in rats and mice, they were regulated by the EPA in a similar fashion as asbestos, aromatic amine dyes, or vinyl chloride, which had actually been shown to cause cancer in people. In the 1990s there were about twenty individual chemicals (including elements) considered by the EPA to be human carcinogens and well over a hundred animal carcinogens. So it was clear that the historical regulatory policies of the government regarding carcinogens were being primarily driven by chemicals that had been identified as animal carcinogens and not by those proven to be human carcinogens.

Much of the regulatory activity in the United States is guided by a philosophical tenet called the precautionary principle, which holds that in the face of uncertainty one should take the most health-protective course. In such general terms, there is nothing unusual about the precautionary principle; it is often invoked in everyday life. Although it is believed to have originated with the German idea of *Vorsorgeprinzip* (foresight) or the Scandinavian "prudent avoidance," I believe that something like it has always been with us.[8] The Hippocratic Oath is an example of the precautionary principle, stating that above all physicians should do no harm. In other words, if you don't know that something will help the patient, don't do it. As applied to the regulation of chemicals, the principle could be paraphrased to say that unless you know toxic effects in animals won't happen in humans, assume that they will.

In recent decades, invocation of the precautionary principle has become steadily more explicit, mainly in legislation intended to prevent environmental degradation. The 1987 Montreal Protocol regulating ozone-depleting substances, the 1992 Climate Change Convention, the 1992 Rio Declaration on Environment and Development, and the 1995 UN Agreement on High Seas Fishing all incorporated the precautionary principle into their texts. It has received its strongest and most widely accepted formulation in the Wingspread Statement, which emerged from a workshop of activists, government scientists, and representatives of various other groups held in 1998: "When an activity raises threats of harm to human health or the environment, precautionary measures should be taken even if some cause and effect relationships are not fully established scientifically."[9]

For toxic effects other than cancer, the EPA performs an in-depth evaluation of other toxic effects, including organ-specific toxicity. The purpose of this toxicological evaluation is to determine what are called "critical adverse effects." These are the noncancer toxic effects that occur at the lowest doses in animal toxicology tests. EPA regulators can base their exposure limits on the basis of the identification of certain noncancer toxic effects at particular doses. The regulatory agencies use this information to determine what "safe" levels of chemicals are. The doses of these critical adverse effects are reduced based on the precautionary principle, using various "uncertainty factors" that can range from a hundred- to ten-thousand-fold depending on the study design. The uncertainty always involves a reduction in the allowed exposure in order to be precautionary.

• • •

The only exception to this method of evaluation of carcinogens occurred when the IARC classification system was changed in 1992 to allow for an upgrade or a downgrade of animal carcinogens based upon cancer mechanism data.[10] The results of genotoxicity tests and similarities or differences in cancer mechanisms between the experimental animals and humans could be used to increase or decrease the suspicion that a chemical could cause cancer in humans. Initially, however, these were used in a lopsided manner, such that the mechanistic data were cited only to upgrade the classification to a human carcinogen, even when the human data was "limited." From 1992 to 1998 twenty-one chemicals were upgraded, and only one was downgraded, even though the rules in the preamble allowed for either.[11] It wasn't that many of these chemicals did not deserve to be upgraded, but there were chemicals not given consideration that should have been downgraded as well. In contrast, the U.S. EPA in 1991 had decided to discount certain male rat kidney tumors in their evaluation of carcinogenicity. This meant that agents such as limonene and gasoline did not need to be classified as carcinogens. So the U.S. EPA essentially downgraded chemicals while the IARC did not.[12]

The lack of an even-handed approach by the IARC was not only a matter of the evaluations of the chemicals by the members of the working group but also of the choice of chemicals evaluated. And who was invited to join the working groups and which chemicals were to be evaluated was the purview of the director of the IARC, which at the time was Lorenzo Tomatis, who retired in 1993. After Tomatis left, one pharmaceutical, clofibrate, was the first to be downgraded. When Douglas McGregor was temporarily in charge of the monographs and before Jerry Rice arrived on the scene, the working group in 1996 for monograph 66 decided that liver tumors caused by clofibrate were not relevant to human cancer.

Paul Kleihues became director of the IARC in 1994, but he had his own active research practice studying brain cancer and other research programs to supervise. Therefore a separate position was created to oversee the production of the monographs. Jerry Rice came from the U.S. National Cancer Institute to become the first head of the monograph program. Jerry Rice was willing to entertain both upgrading and downgrading classifications based on cancer mechanism, and an IARC workshop was held in November 1997 that developed certain criteria for working groups to consider downgrading classification of a chemical. The crucial moment occurred when it was officially decided that increases in particular types of rodent tumors were not necessarily relevant to humans.[13] This involved establishing a mechanism that caused the tumors in rodents, which was either entirely specific to rodents or which would not be expected in humans because of human insensitivity to the effect.

The working group for monograph 73 was specifically looking at the backlog of some chemicals where the tumors produced in animals might not be relevant to human cancer. I was asked to participate in this working group, so off I went to the IARC headquarters in Lyon, France. Lyon is where the Saône and Rhône rivers come together and was the seat of the Roman Empire in France. It is the third-largest city in France at about half a million people. Usually twenty or thirty scientists from around the world are involved in the working groups, plus various IARC staff members, all together for ten days of meetings. To give some understanding of how the IARC organizes these working groups, each member is assigned to one of the four subgroups of the working group depending on their particular area of expertise. These groups represent experts in chemical exposures, epidemiology, animal bioassays, and cancer mechanisms. I was the leader of the cancer mechanisms group, and my colleague from the American Health Foundation, Gordon Hard, led the animal bioassay group. About half of the scientists were from the United States; the others came from Italy, Norway, Denmark, Germany, Austria, Canada, Japan, and the United Kingdom.

As a result of this working group, four chemicals were downgraded based on cancer mechanism data. Based on the studies by Lois Lehman-McKeeman, at the Miami Valley Laboratories of the Procter & Gamble Company, it was agreed that the tumors in male rats caused by the natural limonene of citrus oils were caused by its binding with a protein found not in humans but only in the male rat kidney. Saccharin was another chemical to be downgraded; it had previously been called a carcinogen because of tumors it caused in the bladders of rats. Studies by Sam Cohen, at the University of Nebraska Medical School, had accomplished an impressive group of experiments that showed that the bladder cancers in rats were caused by microcrystals in the urine, which caused chronic irritation. Given the differences in the physiology of rat and human urine, such microcrystals were not possible in humans, and therefore tumors would not occur.

Another working group for monograph 79 was focused on thyroid tumors in rats. The rat and the human have very different responses to thyroid tumor formation from inhibition of thyroid hormone production. One chemical that was downgraded was amitrole, the very same chemical that had caused the cranberry scare and that Rachel Carson had used as an example of a carcinogen that could affect humans. It turns out that thyroid peroxidase, an enzyme that amitrole inhibits at very low levels in rats, in humans requires much higher levels of exposure to the chemical—levels that were physically impossible to achieve. Another downgrade was for sulfamethazine, a drug used to treat bladder infections that also caused inhibition of thyroid peroxidase.

So, in these two working groups some balance had been achieved, with eight downgradings based on mechanistic data. But Lorenzo Tomatis, a former director of the IARC, issued strident objections to this use of mechanistic data to downgrade, even though the classification rules allowed for this very process. Tomatis's accusations were so intense that Paul Kleihues, director of the IARC at the time, barred Tomatis from setting foot in the building—where he still had an office.[14]

Vincent Cogliano took over as the head of the program from 2003 to 2011, and after that, Kurt Straif. Under their leadership many chemicals have been upgraded and none downgraded based on mechanistic data. The precautionary principle reigns supreme again, unchecked by balanced evaluations. According to Cogliano, as of 2008 there were fifty-two agents upgraded in total, compared to the eleven upgraded and eight downgraded during the time that Rice was head of the monograph program. And since 2008, there have been at least eight more agents upgraded and no more downgraded based on mechanistic data.[15]

18 | The Dose Makes the Poison

Harking back to Paracelsus's concept that the dose makes the poison, one medication that is widely used but can be very toxic to the liver is acetaminophen, commonly known as Tylenol. Acetaminophen is an excellent treatment for pain and inflammation, but in children its most important use is to control fever. It is very effective at lowering body temperature. Most people do not recognize that the dose required for the therapeutic benefits of this medication is reasonably close to the dose that is toxic.[1] The therapeutic index of a drug is the ratio of the toxicity of a drug to the effective dose, and for acetaminophen this ratio is not as high as toxicologists would prefer.

After ingestion, chemicals, including drugs, are combined in the liver with highly water-soluble conjugates of either sulfate or glucuronide and then eliminated in the urine. These are two of the most important detoxification pathways, and they rarely create toxic products. However, as we saw in chapter 10 the liver also contains other metabolic pathways: a cytochrome P450 called CYP2E1 also activates acetaminophen to a very toxic product, abbreviated NAPQI, which can cause liver necrosis.[2] As the dose of acetaminophen increases, the detoxification pathways become overwhelmed, and more and more of the drug activated by CYP2E1 produces more and more liver toxicity. In adults, hepatoxicity may occur after ingestion of a single dose of ten grams, which is only ten times more than the usual therapeutic dose of one gram. Twenty grams can be fatal. Some people try to take their lives with Tylenol because it is in the medicine cabinet but end up surviving with a severely damaged liver.[3] In children, there is evidence that twice the recommended dose of acetaminophen given over a few days can result in liver toxicity.[4]

The enzyme CYP2E1 has an older name—alcohol inducible cytochrome P450. The liver has developed the capacity to increase levels of cytochrome P450 types after exposures to some chemicals by increasing gene expression, thereby producing more of the enzymes. The inducer, in this case ethanol, signals the nucleus to increase the production of messenger RNA from the *CYP2E1* gene, which in turn increases the synthesis of the CYP2E1 protein, which causes more enzyme activity. All this would merely be interesting information were it not for the resulting increased liver toxicity by acetaminophen in heavy consumers of ethanol. So in this case, toxicity is not just the result of the dose of the potential poison but is exacerbated by the lifestyle alcohol use of the consumer.[5]

Ethanol consumption has its own dose-related effects. Ethanol is rapidly absorbed from the gastrointestinal tract and metabolized in the liver by alcohol dehydrogenase to acetaldehyde, which is somewhat toxic. But the acetaldehyde is rapidly metabolized to acetic acid, that is, vinegar, which is not very toxic at all. This metabolism eliminates about ten grams of alcohol per hour from the human body, which means that in ninety minutes a person can metabolize the amount of ethanol in a shot of spirits, a glass of wine, or a can of beer. Once one exceeds that rate of consumption, the blood alcohol level rises.[6]

Dose is expressed as the mass of a chemical that is taken into the body, which for a single exposure is measured in grams. It can also be measured in grams per kilogram body weight; this is commonly used in experimental studies. For chronic doses, the metric is grams per kilogram body weight per day, but often milligrams (one-thousandth of a gram) are used instead. The dose of drugs, especially for pediatric medications, is often described in mass per surface area of the body. For pharmacologic or toxic effects, the important measure is the amount of chemical present at the target organ, which in the case of ethanol is the central nervous system or the liver. The target organ dose is not measured in the usual dose metrics but rather in concentration. So for ethanol, the target organ dose is measured in milligrams per 100 milliliters of blood, and because ethanol can cross the blood-brain barrier, blood concentration is used as a surrogate for brain concentration.

The lowest concentrations of ethanol provide the euphoric effects that make alcoholic drinks so desirable. Sobriety from a legal standpoint is usually defined as an ethanol concentration below 80 mg/100 ml. Lack of coordination, slow reaction time, and blurred vision can occur from 50 to 150 mg/100 ml of blood alcohol. Higher levels of 150 to 300/100 ml cause visual impairment, staggering, and slurred speech. At even higher levels of 300 to 500 mg/100 ml, there is stupor, low blood sugar, and convulsions. Once over 500 mg/100 ml, coma and death results in all but the most highly alcohol-tolerant people.[7]

It is clear that a chemical may have several thresholds, each for a different effect. Alcohol causes liver toxicity above a threshold dose, but below that dose there is no evidence of liver damage. There is another threshold—usually lower—above which alcohol elevates mood, while a very small amount has no perceptible effect. At doses below the threshold, the homeostatic, or self-regulatory, mechanisms of the body readjust the effect of a chemical. In fact, low doses of alcohol prevent heart disease and lessen the risk of heart attack. This illustrates that our bodies are constantly exposed to doses of all kinds of "natural" chemicals in our food and drink, but we don't get sick unless we consume too much of the wrong thing.

The same is true for chemical exposures. People are able to tolerate low levels of exposures thanks to elimination of chemicals from the body. Even when a chemical is present in a sufficient enough concentration in the body to have an effect, it can be counterbalanced by other homeostatic mechanisms. If, for example, a chemical lowers blood pressure, receptors in the cardiovascular system detect this and provide signals to the heart to increase its rate and to the blood vessels to contract. These effects raise the blood pressure back to normal levels. Even if DNA is damaged by a chemical, we learned earlier that the tumor suppressor gene produces *p53*, which halts cell replication until either DNA repair can be accomplished or programmed cell death can take place. In other words, a cell either repairs itself or dies quietly without causing an inflammatory response. Homeostatic mechanisms are responsible for creating a threshold dose that must be overcome before an effect is apparent, thereby creating a threshold dose required for an effect.

Another basis for threshold effects is hormesis, which has been defined as a dose-response relationship whereby there is a biological activation at low doses but an inhibition at high doses, or vice versa. This concept was promoted by Edward Calabrese, at the University of Massachusetts.[8] Hormetic effects have been studied for more than three decades, and many toxicants have shown benefits, rather than harm, with low-level exposure.[9] Such effects occur in some instances but not in others and so need to be appreciated on a case-by-case basis. Henry Pitot, at the McArdle Institute of the University of Wisconsin, studied the promotion of precancerous liver lesions by dioxin and found the U-shaped dose-response curve typical of hormesis.[10]

• • •

The liver is particularly prone to damage from ingested chemicals because they go directly from the intestine to the liver via the portal system. The portal vein drains the gastrointestinal tract, including the stomach and intestines, directly to the liver before the blood returns to the heart. This allows the body to have digested nutrients unloaded directly into the liver, the metabolic engine of the body. The kidneys are also susceptible to toxicity for some chemicals because they may be concentrated in the urine.

The central nervous system is protected from some chemical assaults by the blood-brain barrier produced by the particularly tight connections between the cells surrounding the blood vessels. For example, venom components are large protein molecules that can't get past the blood-brain barrier. As a result, venoms do not affect the central nervous system but have their effect on the peripheral nervous system. Chemicals such as alcohol, however, have easy access to the brain because of their ability to pass through cell membranes. Thus too many drinks go straight to altering mental activity. In contrast, the components of the peripheral nervous system, for example the neuromuscular junction, are not protected by any barrier and so are susceptible to chemicals such as venoms.

The skin and lungs are in direct contact with the air in the environment. For occupational exposures the skin and lungs account for one-quarter of all nonfatal occupational illnesses, with the skin contributing a larger share. Many of these illnesses are caused by direct contact with industrial chemicals or natural substances.[11] The dose involved in the contact may be relevant to injuries such as chemical burns, but a major cause of occupational skin disease involves allergic reactions where dose is less important than individual sensitivity and a history of previous exposures.

The lung is the organ most likely to experience effects from toxic substances in the air. In general, these are not chemicals that require metabolism but ones that react directly with cell membranes or cause an inflammatory response. Ozone is a chemical that is toxic to cell membranes and can be present in large quantities in the air, mostly in urban environments as smog. At higher concentrations, it causes toxic effects and destroys the alveolar air sacs where oxygen is exchanged for carbon dioxide. Sulfur dioxide, another chemical present in urban air, causes irritation in the air passages, producing constriction, so that it is difficult to exhale, as in asthma. Other agents, those implicated in the pneumoconioses, such as coal and asbestos, cause fibrosis of the lungs, preventing them from properly expanding and contracting.

Besides direct effects in the lung, inhaled chemicals can cause systemic toxicity after their absorption into the blood and subsequent circulation to organs and tissues. In this case, an exposure level in the air becomes a concentration in the blood. Not all of the chemical is usually absorbed, but because of the ability of some chemicals to cross cell membranes, absorption rates can come close to 100 percent. Therefore, for some chemicals air concentration can be correlated to blood concentration.

The major exceptions to the traditional dose-response effect are those toxic reactions mediated by the immune system. Chemicals that cause sensitization in some people can produce effects at very low doses in subsequent exposures. For people who are resistant to these effects, subsequent low exposure will have no effect. Occupational exposures to beryllium was described in a previous chapter, whose toxicity at lower levels is due to an immune response. Allergic contact dermatitis is a common dermatological disease that is immune mediated and develops in two sequential phases: induction and elicitation. In the induction phase, topical exposure leads to sensitization of the individual. Later exposures cause the toxic effect in the form of a rash. Topical antibiotics, rubber products, antiseptics, and metals are among the most common causes of contact dermatitis. High-risk occupations are in health care, hairdressing, the food sector, and the metal industry. Thousands of different chemicals have been implicated, and their effects can have serious consequences for consumers and workers. One of the most common causes of this illness is the nickel used in body piercings. In health care workers, latex gloves are a common cause of contact dermatitis.

Isocyanates, for example, are one of the principal reagents used in the manufacture of polyurethane plastics, and they can be extremely toxic at low doses. Dr. Reinl, in Germany, described a man who developed asthma in 1944 while working with toluene diisocyanate. He died in 1952 from complications from chronic emphysema. Subsequent investigators described the lack of consistency in the presentation of respiratory symptoms: some workers became ill within an hour of going into the plant, others did not develop symptoms until some hours after work, and others not at all. The type of onset and exquisite sensitivity was suggestive of allergic asthma, and an immunological mechanism was found for this disorder.[12]

• • •

Yet another dose-response relationship has been described for chemicals that cause cancer. The cumulative dose of a chemical is the important determinant of the development of cancer caused by chemical exposures. Although bone marrow toxicity can be caused by very high levels of benzene in the air over a short duration, cancer development requires exposure over time. In a 1930s study of rubber workers who developed severe bone marrow toxicity including aplastic anemia, the air levels rose as high as 500 ppm over short periods.[13] In the study of the Pliofilm workers published in the 1980s, levels generally were lower, but the exposures took place over longer periods of time.[14] For leukemia risk, rather than just the intensity of exposure being important, the cumulative exposure is the determining factor, which includes both the intensity and the duration of exposure. This cumulative dose is measured as the average benzene air level multiplied by the years of exposure.[15]

This dose-response information had two important aspects. First, there was a level of benzene exposure that did not exhibit an increased risk for leukemia. Thus, for benzene exposure and the mechanism that causes leukemia, a homeostatic mechanism seems to provide protection at low doses. If indeed topoisomerase II inhibition is the mechanism, then a certain amount would have to be inhibited for the genotoxicity to occur, since there appears to be an excess of it. Also, since the production of benzene metabolites is also an important step in genotoxicity, then a certain amount of them must be present for cancer development. This means that the dose-response curve would have a threshold leading to a sublinear shape, which means that the slope is constantly increasing. The reason for the increasing slope again has to do with the increasing dose overcoming the body's protective measures. As the dose increases, the protective mechanisms become more and more overwhelmed. Unfortunately, there are no animal studies to test this dose response because rats and mice do not get leukemia from benzene exposures.

Another example of cancer dose-response has been developed for asbestos exposures in industrial settings. Asbestos has been used in a broad variety of industrial applications, drawing upon its low cost and desirable properties such as heat insulation, fire resistance, and ability to withstand friction. At the peak of its demand, there were about three thousand applications or types of products using asbestos. In most of its applications, asbestos is bonded with other materials, such as Portland cement, plastics, and resins. In other applications, asbestos is used as a loose fibrous mixture or woven as a textile.[16] The shipbuilding industry uses many different asbestos products, mainly in pipe lagging and in lining rooms with asbestos-containing boards. Men of various trades work together in confined spaces, sharing each other's occupational hazards.[17]

Richard Doll evaluated the studies of lung cancer associated with cigarette smoking and noticed that a large number of cases of lung cancer were found in persons with asbestosis. He noted that given the infrequency of asbestosis, this large percentage of cancers suggested—but did not prove—that lung cancer was an occupational hazard for asbestos workers.[18] There was previous evidence from the number of studies conducted in several different countries that miners with a history of exposure to asbestos dust were at risk of developing asbestosis and perhaps lung cancer.

Irving Selikoff, of Mount Sinai Medical School, and E. Cuyler Hammond, of the American Cancer Society, began an investigation in 1966 to obtain information on the combined effects of cigarette smoking and exposure to asbestos dust with respect to death from lung cancer and chronic noninfectious pulmonary disease. Their study presented conclusive evidence of an increased bronchial carcinoma risk in U.S. and Canadian asbestos insulation workers who were members of the International Association of Heat and Frost Insulators and Asbestos Workers. They showed that asbestos increased the risk fivefold, that smoking increased the risk more than tenfold, and that the combination of asbestos exposure and smoking over fiftyfold.[19]

This relationship is considered to be multiplicative or synergistic. In toxicology, synergy is a relationship between two causes of a disease whereby they have a greater than additive effect. Mathematically, rather than the fivefold increase in lung cancer caused by cigarette smoking adding to the tenfold increase caused by asbestos resulting in a fifteen-fold increase, the increase is fiftyfold. The epidemiologic evidence from other studies of insulation workers with high exposures also supported the synergistic model, indicating that each of these two factors may have an independent action on the multistage process of carcinogenesis.[20] Other potential mechanisms for this synergy are that carcinogens in tobacco smoke may be adsorbed (that is, form a thin film) on the surface of asbestos fibers, thereby increasing the uptake and retention time of these carcinogens or the penetration of target cells by asbestos.[21]

The other cancer caused by asbestos was especially pronounced in the shipbuilding industry during World War II. Because this disease has a long latency—it usually takes more than thirty years to develop—mesothelioma was identified and studied later than lung cancer in industrial settings. For example, the number of deaths from mesothelioma increased in the United

Kingdom from about fifty per year in 1968 to 1,600 per year in 2001.[22] In contrast to lung cancer, smoking does not cause mesothelioma, nor does it have a synergistic effect with asbestos exposure.[23] Eventually mesothelioma was found to result from asbestos exposures in many industries, including insulation, insulation, construction, and asbestos production, and from living in proximity to an asbestos production site.[24]

19

Are We Ready to Clean Up the Mess?

Risk assessment is a mathematical approach to the question of whether people can get cancer or suffer other health effects from exposures to chemicals. Previously, the answer to this question was only in qualitative rather than quantitative descriptions. It was based upon the similarity of chemical-exposure circumstances in which observations had been made. An example for cancer would be the benzene air levels found in the Pliofilm workers that increased the risk of acute myelogenous leukemia. If another group of workers were exposed to similar benzene air levels over similar time periods, one could predict that a certain number of excess acute myelogenous leukemias would occur. But the extrapolation of cancer risks associated with much lower exposures in environmental situations from measured risks in high occupational exposures or rodent bioassays had yet to be developed.

Ken Chase, on the faculty at George Washington University, started a medical practice and consulting firm focusing on occupational medicine called Washington Occupational Health Associates. In one of his consultations, Ken served as the medical director for Amtrak, the national railroad passenger train company. Amtrak's biggest environmental problem would become PCBs, which were used in its electrified locomotives and passenger cars on the East Coast. Based on his examinations of the possible health effects among Amtrak employees who had been exposed to PCBs, Chase published an article in the *Journal of Occupational Medicine* in 1982, where he reported that the only detectable effect of PCBs in these workers was an association between blood PCB and lipid levels. However, in retrospect, this turned out to be more complicated in that the cause-and-effect relationship was the opposite of that reported. Because PCBs are lipid soluble, they are concentrated in the fat tissues. But in people who had higher blood lipid levels, more PCBs were partitioned into the blood, resulting in higher blood concentration. This is a classical problem for the relationship between cause and effect in statistical associations.[1]

Both the Detroit Edison Company and Commonwealth Edison of Chicago contracted Chase in 1984 to help them figure out how to approach the cleanup of PCBs in soils. PCBs were used in the Edison Company's transformers and capacitors. In the case of Chicago Commonwealth Edison, the problem was that their pole-mounted capacitors would explode or leak, sometimes in residential neighborhoods. One to 2 percent of these units per year were spraying PCBs onto the surrounding soil and other surfaces.

I had known Ken for about ten years, and he knew about my toxicology work in childhood lead poisoning, especially the quantitation of lead exposures and disease. I began consulting for Ken on these projects involving the development of PCB exposure calculations for risk assessment to find cleanup levels for soils that would be health protective. Although we didn't know it at the time, Commonwealth Edison was developing its cleanup levels for PCBs because of a court-determined consent decree with EPA's Region 5 office. Many aspects of our methodology for sampling soils and developing PCB risk levels in the environment were subsequently used by the EPA in mandated soil cleanup at other sites. The core of the problem of cleaning up the mess was to determine what constituted an acceptable level of cleanup: what concentration of PCBs in soils was dangerous, and what concentration is small enough to be considered safe? This is a particularly important question considering that it would be physically impossible to remove every molecule of PCBs, or any other chemical for that matter, from an industrial site. In other words: how clean is "clean"? The determination of acceptable levels of chemicals in the environment relies on a methodology called risk assessment, which had been only rudimentarily developed in an ad hoc mode by the early 1980s. There had been some recommendations for risk assessment developed by a committee of the National Academy of Sciences in 1984. But the major difficulty was in exposure assessment, which is often difficult to estimate accurately and for which there was little guidance at the time.

My experience with the lead poisoning issue eight years earlier was helpful as a start. I had tried to figure out how much lead paint a child might ingest and how the lead that they ate could cause a certain blood lead level. But the exposure assessment for PCBs was much more complicated. Also, this was a cancer risk determination that needed to be made based upon PCB carcinogenicity studies in animals. Human epidemiology studies were available, but they did not show any consistent evidence of cancer risk, so we relied on rat studies for our evaluation.

The science of risk assessment was primitive in the early years; quantitative information was often lacking for chemical exposures in humans and for cancer development. Consequently, scientists developed risk assessment methodology using animal studies to try to quantify the possibility of risks and at least come up with an upper limit on what the risks might be in humans. In animal bioassays, the doses of chemicals could be controlled, and incidences of cancer could be readily determined. But one had to assume that at least an upper limit on human cancer risk could be estimated from these rodent bioassays.

Unfortunately, at the time there was no experiment with PCBs that gave us dose-response information in either humans or animals. There were several animal studies that found an increase in tumors from just a single dose of PCBs. One study done by the National Cancer Institute using three doses showed some tumors, but none were considered to be statistically significant increases. Having a study with a single dose didn't stop the EPA from putting together a dose-response estimation that would be used for cancer risks for human exposures. It was called a "cancer potency factor" that when multiplied by the estimated human exposure gave the cancer risk for humans.[2]

And in line with the precautionary principle described previously, the U.S. regulatory agencies assume that there is *no* safe dose for chemical carcinogens. This was in contrast with other types of toxic effects, where these same agencies assume that at a low enough dose there will be no toxic effect. It was also in contrast to some countries, such as the Netherlands and the United Kingdom, who took an approach very different from the U.S. EPA, whereby they use the threshold dose concept for nongenotoxic animal carcinogens. Other governments evaluate chemicals on a case-by-case basis and do not have hard-and-fast rules governing the classification of animal carcinogens or low-dose considerations.[3]

Because this dose-response estimation was based on such shaky information, it was designed to be protective and adhered to the precautionary principle, so that it only gave an upper limit on the possible risk. Critically, there were two large assumptions included in this risk estimate. The first was that the finding of liver tumors in rats was predictive of human cancer. The second was that the extrapolation from the very high doses used in these rats could predict a risk of human cancer at a much lower dose. My job for the electric utilities was to come up with cancer risk calculations using this cancer potency factor and exposure estimates from the PCBs in soil. For example, if the soil contained ten parts per million of PCBs, which is the same as 0.00001 percent of the total weight of the soil or sediment that contains it, what quantity of PCBs would a person take into their body each day, and what would be their resulting upper-bound cancer risk?

First, we had to figure out the pathways of exposures: inhalation, ingestion, or dermal. In order to come up with an acceptable level of PCBs in soils and surfaces under various hypothetical circumstances, I had to do twenty-three different exposure and risk calculations. These included dermal absorption for preschool children, school-aged children, and adults, taking into consideration various indoor and outdoor activities. For oral exposures we calculated risks for soil ingestion, consumption of garden products, animal fat and milk for animals grown on the premises, and even consumption of wild game. For inhalation we calculated risks associated with breathing PCB vapors and PCBs bound to particles.

We hoped that we were covering all the possible bases since we doing this for the first time.

• • •

General Electric was one of two companies on the hook for PCB contamination because they manufactured most of the electrical equipment containing them. The other was Monsanto, who had produced the majority of PCBs in the United States from 1930 up until 1977, when they were banned from use as insulation in electrical equipment. PCBs are made of two benzene rings attached together with up to twelve chlorines on each of the twelve carbons. There are 109 possible PCBs, with various specific chemical configurations of chlorines, but in practice there were only several mixtures of the types that were marketed by Monsanto. Monsanto's brand of PCBs was called Aroclor, and different Aroclor products contain various numbers of chlorines attached to the carbons.

In order to improve the cancer risk calculations, GE developed an ambitious bioassay plan, testing all four of the major different Monsanto Aroclor types, which accounted for 92 percent of PCB sales in the United States from 1958 to 1977. In this study, 1,300 male and female rats were dosed over their lifetimes with different doses of several Aroclor types with different degrees of chlorination. This study would form the basis of understanding not only what doses of PCBs cause cancer in animals but also provide cancer mechanism information that would explain the development of liver cancer in the animal model system.[4] After the initial dose-response study, some tissues were saved for mechanistic studies, which my lab group performed at the American Health Foundation, resulting in a publication.[5]

This study showed that only the high-dose group of the Aroclor mixture with the most chlorine caused an increase in tumors in the male rats, but for females, which was the sex originally tested by Kimbrough, there were increases in almost all exposed animals. The results of this test provided a more precise and significantly lower and less conservative estimate of the cancer potency for use in risk assessment than the one I had used for the

Detroit and Chicago utility companies. Also, it provided some interesting observations on possible mechanisms of liver tumor formation.

During the 1980s, risk assessment methods continued to be developed, culminating in 1989 with the EPA's Risk Assessment Guidelines for Superfund (RAGS), used to clean contaminated sites. This was essentially a cookbook about how one should perform a risk assessment based upon mathematical information[6] from scientific studies using the precautionary principle. The number of chemicals that were regulated based on cancer risk assessments grew, specifying the allowable amounts of chemicals in drinking water, air, and foods. If it was calculated based on animal studies that a chemical caused possibly greater than a one-per-million cancer risk when it was consumed in drinking water at a certain amount, it would have to be reduced or eliminated.

Although we have seen in the previous chapter that toxic effects usually have a threshold dose, the model used by the EPA for risk assessment assumed that at the lower doses there was no threshold; that is, there was always a minuscule risk of cancer at minuscule doses. If a lifetime increase in cancer risk from the PCB exposure was calculated to be one per million, this is a level of risk considered to be acceptable for widespread exposures in the general population, such as for contamination of drinking water. For contaminations confined to smaller numbers of people, higher risks, up to one per ten thousand, can be acceptable. But the end result was that we predicted risks so small that they were close to nothing compared to background cancer risks. And this becomes the dilemma in cancer risk assessment. The risks calculated become so vanishingly small that they are difficult to comprehend. This is especially true because they are hypothetical, and many seemingly unrealistic assumptions are used to calculate these risks compared to the more than one in three chance of developing cancer in a person's lifetime.

The question of what is an acceptable cancer risk is an interesting one. How did we wind up with one in a million being an acceptable risk? Risk assessment methods did not begin with the EPA but with the FDA. The "DES Proviso" of the drug amendments to the Delaney Clause of 1962 prohibited carcinogenic drug residues that could be detected by analytic methods approved by the FDA.[6] Under the DES Proviso, the FDA could approve a carcinogen for food animal use if the concentration of any residue remaining in the edible tissues was so low that it could not be detected, thinking it then presented an insignificant risk of cancer to consumers. Standard tests at that time were not very sophisticated, and regulators were concerned that drug residues in foods might exist in dangerous levels even if they couldn't be measured with existing technology. Using advances in analytical chemistry, FDA scientists developed techniques to measure more minute concentrations of residues. Defining an acceptable risk level was tied to the increasing ability to measure these drug contaminants in food.[7] So, "clean" got cleaner.

In 1973 the FDA had adopted a one in one hundred million risk level, but no current analytical method could test contaminants at concentrations corresponding to this risk level. This rigorous standard was adopted based on a publication by Nathan Mantel and W. Hay Bryan, from the National Cancer Institute. However, they recognized that it would take 460 million tumor-free animals in order to prove this level of risk at a 99 percent assurance level.[8] (This was before the development of the notion of treating animals at high chemical levels and mathematically extrapolating to low levels.) The vanishingly small risks were statistically associated with impossible-to-measure chemical amounts. To quote Lewis Carroll from *Through the Looking-Glass*:

> "I can see nobody on the road," said Alice.
> "I only wish I had such eyes," the King remarked in a fretful tone. "To be able to see Nobody! And at that distance too! Why, it's as much as I can do to see real people, by this light!"

Apparently, the choice of an acceptable risk level of one in a million was first made by Donald Kennedy, the FDA commissioner, in 1977 after a public hearing and comments on the acceptable risk level.[9] The development of analytical techniques led to a 1987 regulation called the sensitivity-of-method procedures, or SOM. The regulation defines an insignificant risk of cancer as a one in 1 million increase in risk and spells out how to measure the residue concentration. The actual origin of this as an acceptable risk level appears to be arbitrary and difficult to trace.[10]

Risk levels were later adopted by the EPA, but there is little available scientific rationale to understand why acceptable risk levels are what they are, other than that the EPA just adopted the levels of acceptable risk formulated by the FDA. In the EPA's Superfund regulations, cancer risks of as high as one per ten thousand can be acceptable. So why do we sometimes set standards at a less than one in a million cancer risk? According to EPA documents, both one in a million and one in one hundred thousand can be acceptable risks for the general population. For smaller populations, the level of one in ten thousand is acceptable.[11]

Trying to talk to the public about cancer risks is a tricky business and one of the subjects in a discipline called "risk communication." A cancer risk of one in a million is a hard metric to explain, especially when people will spend money to buy a lottery ticket where the chances of winning are less than one in a million. So these are clearly not insignificant odds for some. A risk communicator could use the argument that this is only an upper limit of the risk and that it could be between one and a million and zero, or that 40 percent of people get cancer anyway, and so the increase in risk amounts to only 40.001 percent. Comparing the increased cancer risks to known risk hazards or accident rates does not satisfy an audience of anxious people. The audience is usually focused on the mathematical calculation of risk and sees it as real rather than hypothetical.

• • •

A risk assessment question that we tackled at the American Health Foundation in 1992 was whether the relatively low level of asbestos in public buildings, including schools, was a public health risk. In order to do this, I put together a panel of experts in risk communication, lung pathology, pharmacology, toxicology, and epidemiology. The cast of characters included Vincent Covello, from the Center for Risk Communication at Columbia University, who had written extensively about the best ways to discuss risks with the public in either a publication or public meeting. Marvin Kushner was at the Department of Pathology of the State University of New York at Stony Brook. Marvin, at the time, was quite the elder statesman and previously dean of the medical school at Stony Brook from its inception in 1972 until 1987. Arleen Rifkind was in the Pharmacology Department at Cornell University; I had met her through our mutual interest in dioxin. Karl Rozman was from the Department of Pharmacology, Toxicology, and Therapeutics at the University of Kansas Medical Center. Finally, besides Gary Williams and myself on the panel was Dimitrios Trichopoulos, chair of epidemiology at the Harvard School of Public Health and a good friend of Ernst Wynder.

The purpose of the panel was to determine if the lung cancer risks were significantly associated with measured levels of asbestos in public buildings and whether the EPA's policy of asbestos containment was the correct course of action in public buildings. At the time it was estimated that there were about five hundred thousand commercial, nonresidential buildings that contained friable asbestos-containing materials. The EPA's approach was called "Operations and Maintenance," whereby the asbestos was identified and maintained so that it wouldn't become airborne. The alternative to containment was complete removal of the asbestos; some studies had shown that the abatement procedures actually increased the levels of airborne asbestos. Furthermore, the 1990 costs of complete asbestos removal across the identified buildings could reach $150 billion.[12]

In contrast to the use of animal studies for risk assessment, which was done for most chemicals, such as PCBs, by this time there were several studies of asbestos-related lung cancer and mesothelioma that had good measurements of asbestos levels. Thus the cancer potency factors could be derived directly from epidemiology studies.[13] We concluded that there would be about two extra cases of lung cancer and mesothelioma per year from asbestos exposures in all public buildings. This estimate was based upon extensive published surveys of air levels in these buildings and the assumption that we could extrapolate high-dose cancer studies to very low doses without any threshold dose. Some other estimates were higher, using different assumptions, but our conclusion was that the overall contribution of asbestos in public buildings to lung cancer and mesothelioma would be less than 0.01 percent compared to all lung cancers.

In contrast to the status quo, asbestos abatement procedures increase the risks by causing a spike in the asbestos air levels during the required demolitions. However, this didn't stop New York City's government from digging out all of the asbestos from all of its school buildings in the summer of 1993. I tried to counsel caution by getting in touch with the media and getting permission from *Preventive Medicine*, the publisher of the panel's article, to allow me to discuss our findings before publication. I even went on a local television news program along with Schools Chancellor Ramon C. Cortines to try to convince him and the public that the EPA's maintenance procedures should be used instead of the potentially dangerous abatements. I also pointed out that the estimated cost of abatement of more than $100 million could provide each New York City school with more teachers or health counselors—but to no avail.

Because of the abatement, which was supposed to be completed during the summer school break, of the city's one million schoolchildren, an estimated sixty thousand were either out of school or in different schools—a number that, alone, would form a respectable city school district, one about the size as San Francisco's.[14] There were 115 schools closed and hundreds more disrupted, costing more than $119 million. The last New York City school to be reopened after the asbestos cleanup was in mid-November 1993. According to the *New York Times*, many city and school officials suggested that Mayor Dinkins and other officials had overreacted earlier in 1993 to the asbestos revelations, converting a problem that might have been resolved with a calmer inspection and cleanup effort, carried out over a period of months, into a disruptive crisis.[15] I remember seeing a television program showing teachers sweeping up their classrooms before the children arrived, and they weren't wearing any masks. Even with the vanishingly small risks that we found for asbestos in public schools, it didn't guide the decision-making process. The erroneous decision to remove the asbestos was made based on scare tactics and political pressure, and the science took a back seat to panic.

20 | Legal Battles

When all else fails to rein in toxic exposures to workers, patients, or the public, the legal system is invoked; people sue for damages to health or property. Then it is up to the judicial system to decide whether a person has been harmed or needs to be medically monitored for symptoms of harm in the future. This field of the law is called toxic tort litigation, which is a particular type of personal injury lawsuit where it is claimed that exposure to a chemical has caused or may cause in the future a plaintiff's injury or disease.

To prove or disprove a case, toxicologists, epidemiologists, and physicians that specialize in the disease involved in the case are often used as expert witnesses. Expert testimony has always posed a problem because of the expert's possible bias, and the legal system has struggled to develop standards and guidelines for those who function in that role. In 1353, surgeons were called to testify on the question of whether a wound constituted a "mayhem," that is, an injury caused by a violent act. At that time, these experts were regarded as assistants of the court, but by the seventeenth century, they were being treated as witnesses representing one side or the other.[1]

The underpinnings of expert witness testimony are found in the academic discipline of forensic toxicology. Mateu Orfila (1787–1853) was an outstanding figure in European medical circles, and his work had a profound effect on molding the public perception of forensic medicine long before he served as an expert witness. He was dean of the Paris Medical Faculty, a founding member of the Academy of Medicine, the author of several reference books on toxicology, the editor of influential medical journals, and an expert witness in many popular poisoning trials. While lecturing in April 1813, he obtained the precipitates that characterize arsenic in front of his students and affirmed categorically that the same result could be obtained when the poison was mixed with organic fluids, broth, or drinks like coffee or wine.[2]

Orfila performed many experiments with dogs, varying the quantity of poison and the route of administration, testing antidotes and compiling clinical data on symptoms of poisoning and the anatomical damage found in human autopsies. His major work, *A Popular Treatise on the Remedies to Be Employed in Cases of Poisoning and Apparent Death*, was published in 1818. Similar to the teachings of Paracelsus regarding the contemporary practice of medicine and toxicology, Orfila provided many examples of "useless or dangerous" antidotes for arsenic, lead, and vegetable poisons, and he both suggested new antidotes and revisited old ones whose effectiveness, he claimed, had been proven.[3] Orfila served as an expert witness in several famous legal proceedings, and using his own improvements on the arsenic detection methods, Orfila helped to uncover the truth about the murders of Nicolas Mercier in 1838 and Charles LaFarge in 1840. However, Orfila wished to avoid controversy and refused to participate as an expert witness after 1843.[4]

Toxicologists and epidemiologists have testified in countless cases, but expert testimony can often be confusing to judges and juries. In 1858 the U.S. Supreme Court observed that "experience has shown that opposite opinions of persons professing to be experts may be obtained to any amount." Somewhat cynically, the court observed that some of the testimony involving cross-examination of experts was useless, "wasting the time and wearying the patience of both court and jury, and perplexing, instead of elucidating the questions involved." This led to the limitation of the amount of time spent cross-examining such witnesses. Eventually, the Supreme Court ruled to examine expert witness testimony more closely in the 1923 *Frye v. United States* trial decision. This was a case involving testimony using lie detectors, and this decision ruled that expert testimony was admissible only if it incorporated principles and methods generally accepted by the relevant scientific community. However, this decision did not define how one determines what is good, generally acceptable science.[5]

• • •

It took another fifty years before the judicial system would delve into the scientific principles involved in the testimony of an expert witness. The drug Bendectin was marketed beginning in 1956 as an over-the-counter drug that was useful for assuaging the nausea and vomiting experienced during pregnancy. It was a combination of the antihistamine doxylamine and vitamin B6; as both had been on the market previously, extensive testing of Bendectin was not initially done. Questions about its safety began to arise in 1969 based on case reports of babies with birth defects born to women who had taken the drug, and in 1977 the first lawsuit was filed. Things really got rolling in October 1979, when the *National Enquirer* reported: "Untold thousands of babies are being born with hideous birth defects. Two infants are born without eyeballs. Another without a brain. . . . It's a monstrous scandal that could be far larger than the thalidomide horror."[6] In 1983 Merrell Dow Pharmaceuticals responded to the growing wave of lawsuits by withdrawing the drug from the market despite the fact that it had won all but about three of the cases that went to trial. By 1994 more than 2,100 lawsuits had been filed against Merrell.[7]

In 1984 two of the women who had taken the drug Bendectin to combat morning sickness gave birth to children with severe defects. *Daubert v. Merrell Dow Pharmaceuticals* was brought by William Daubert, the husband of one of the two women, along with other members of the two families. The trial judge examined proposed evidence from nine experts and ruled that only Merrell's expert could testify. This physician-epidemiologist had reviewed dozens of epidemiologic studies of Bendectin use and health effects in large groups of women, and he concluded that the data didn't support a link between the drug and birth defects. Plaintiffs' experts had intended to use animal data and comparisons of the chemical structure of the drug with that of agents that cause fetal harm as well as an unpublished reanalysis of epidemiologic studies to show that the drug might cause birth defects. However, the judge decided that because a wealth of published epidemiological data on some 130,000 women existed, admitting any evidence other than the published epidemiology studies was unjustified. The plaintiffs appealed to the Supreme Court, which affirmed the lower court's decision in 1993.[8]

The important distinction between *Frye* and *Daubert* was that the former only included a general acceptance test of the methods used by the expert but not the substance of the expert's opinion. In the Supreme Court's decision in *Daubert*, the judges would be required to evaluate the scientific evidence and determine the scientific merit of the evidence proffered by experts. Also, *Daubert* asks whether the evidence fits the issue at dispute in the case. For example, a scientist proves that fruit fly DNA in test tubes will develop a mutation from exposure to a chemical. *Daubert* asks whether that finding has relevance to the development of a cancer in a particular person.[9]

As a consequence of the *Daubert* decision, the Supreme Court directed trial judges to become more proactive in culling unreliable or less-than-compelling scientific testimony from cases they oversee.[10] The majority opinion, written by Justice Blackmun, highlighted the notion that the differences between science and law necessitate a more selective admissibility standard. Blackmun noted that "there are important differences between the quest for truth in the courtroom and the quest for truth in the laboratory. Scientific conclusions are subject to perpetual revision. Law, on the other hand, must resolve disputes finally and quickly." Blackmun argued that it may be useful to consider a wide range of information in the scientific process but not in the legal process. Given the vastly different objectives of science and law, some form of adaptation is necessary to make scientific evidence useful in the legal realm.[11]

One of the eventual tests of the *Daubert* case was also related to the female reproductive system: silicone breast implants. Although surgical reduction in breast tissue had been previously used, the first known attempt at breast enhancement was in 1895 in Germany, using a benign fat tumor that had occurred on a woman's back. Other attempts followed, experimenting with various substances, and finally silicone, formed from the elements silicon and oxygen, was found to be the superior product. In 1961, after it became apparent that injecting silicone directly into the breast led to problems such as infection and inflammation, the plastic surgeons Thomas Cronin and Frank Gerow, of Houston, Texas, developed the breast implant with the help of Dow Corning. Placement of these plastic envelopes of silicone proved to be relatively simple, and they felt more like natural breasts than any other type of material that had been tried. By 1992 the number of women with breast implants in the United States had risen to about one million.[12]

A decade earlier, in 1982, an Australian physician reported an autoimmune connective tissue disease in three women who had silicone-filled breast implants. Autoimmune diseases include rheumatoid arthritis, systemic lupus erythematosus, scleroderma, and Sjögren's syndrome. The use of these breast implants was initially not within the FDA's regulatory purview, so the usual safety data for medical devices was not available. Beginning in 1984, lawsuits were filed in Australia, followed by others in the United States. Meanwhile, Ralph Nader's consumer group Public Citizen became involved and petitioned the FDA in 1988 to ban the use of breast implants. The sensational message of silicone breast implants as dangerous devices foisted on unsuspecting women was delivered by the reporter Connie Chung in 1990 on her television program. She interviewed women who claimed that they had an autoimmune disease caused by their breast implants and implied that the FDA was to blame for allowing risky products on the market. And when David Kessler took over the position of commissioner of the FDA in 1991, he decided that he must take some action. In 1992 a ban on silicone breast implants was issued, with an exception for women in breast cancer reconstruction research studies.[13]

In the first of the original Australian cases, that of Maria Stern, the jury awarded $2 million in 1984. Many other cases were filed and went to trial, and, in 1991, a federal jury awarded Mariann Hopkins $7.34 million. Following the FDA ban in 1992 came a tsunami of lawsuits: 16,000 within the next two years. Dow-Corning had withdrawn their implants from the market, and there only remained two other manufacturers: Mentor Corporation and McGhan.[14]

The scientific evidence for these autoimmune disorders related to silicone breast implants was slower in coming than the lawsuits. The first known case was that of a fifty-two-year-old Japanese woman who underwent breast augmentation with silicone injections in 1958. She was in good health until 1974, when she developed a dry mouth. Three years later she developed swollen stiff fingers, joint pain, Raynaud's phenomenon in winter, and eventually scleroderma.[15] In 1982 the first case series describing autoimmune disorders following augmentation mammoplasty with gel-filled prostheses was reported. By the time of the FDA hearings, there were 120 case reports published through 1991 used to support an association between autoimmune disease and breast implants.[16]

In 1994 the first epidemiological study examining the risk of autoimmune disorders and silicone breast implants was published by investigators at the Mayo Clinic.[17] By the following year, seven controlled studies had been reported, each of which provided a quantitative assessment of the risk of connective tissue diseases among women with breast implants. All seven studies failed to demonstrate an excess risk for any of the autoimmune conditions among women with breast implants.[18]

Eventually more than four hundred thousand cases were filed in federal and state courts alleging injuries arising from leakage or rupture of the shell that encased the silicone gel. Given the evolving research concerning the extent to which the silicone gel caused or exacerbated connective tissue diseases or immune system dysfunction, these cases represented a challenge for the courts. Besides toxicology and epidemiology, assessment of this research required an understanding of several other areas of science, including immunology, rheumatology, chemistry, and statistics.

Among those early cases was a group of approximately seventy cases tried in *Oregon Hall v. Baxter Healthcare Corp.* Judge Jones, who presided over these cases, used an expert panel to help resolve twenty-five joint motions to exclude the testimony of plaintiffs' expert witnesses in cases consolidated for trial. Because Judge Jones was considering the admissibility of challenged evidence, he framed his instructions to the panel of experts around the *Daubert* standards. Judge Jones appointed four technical advisors: Merwyn R. Greenlick for epidemiology, Robert F. Wilkens for rheumatology, Mary Stenzel-Poore for immunology and toxicology, and Ronald McClard for biochemistry. Judge Jones had screened all four for signs of potential bias and appointed them experts to assist him in determining whether the parties' expert testimony rested on reliable scientific methodology.[19]

On December 26, 1996, Judge Robert E. Jones made legal history when he excluded all testimony by the plaintiff's experts to the effect that silicone-gel breast implants cause autoimmune system disorders. The court reached this result largely based on the four independent experts advising it on the state of scientific knowledge.[20]

Besides the use of the *Daubert* decision, the Bendectin and silicone breast implant stories have one more thing in common: the role of coincidence rather than causality. In the case of Bendectin and birth defects, it is important to understand that, in 1983, it was estimated that 33 million pregnant women throughout the world had used the drug to curb their nausea and vomiting. It was the only drug approved in the United States for treating morning sickness, and in 1982 one in ten pregnant U.S. women used it. It was also estimated that about 5 percent of babies born in the United States each year have some type of birth defect, and about half of those have serious abnormalities.[21] Using simple math, because of coincidence alone there should be almost a million children born with serious birth defects whose mothers took Bendectin.

In the case of silicone breast implants, a large number of women had them, and the autoimmune diseases occur in the same age range when many women want or need breast implants. A study from the Mayo Clinic found that over 8 percent of women will get one of the inflammatory auto-immune rheumatic diseases included in the silicone breast implant litigations.[22] Again, doing the simple math, by coincidence, eighty thousand women with silicone breast implants could wind up with one of these diseases. So when such conditions are not rare, one must be very careful to draw conclusions from occurrence alone, even when the disease follows shortly after the use of a drug, device, or chemical exposure.

• • •

The next question to be addressed by the courts was the use of expert testimony in environmental lawsuits. This occurred in a case concerning environmental contamination in a residential neighborhood. Electrification of the railroads developed in the twentieth century. The Philadelphia commuter system, the Southeastern Pennsylvania Transportation Authority (SEPTA), and eventually Amtrak were electrified on the Northeast Corridor lines. The plaintiffs in this case lived for many years in the vicinity of the Paoli Pennsylvania Railroad Yard, which was a railcar maintenance facility for over a quarter century. Because PCBs were fire resistant, they were used in the large electric transformer that powered the electric engine and the air conditioners of the rail cars. From the required maintenance of the equipment, PCB contamination was widespread on the maintenance yard, and some of the soils eventually migrated into the surrounding residential neighborhood.

The plaintiffs in the case lived in the neighborhood for many years and in 1986 sued SEPTA, Amtrak, and Penn Central, who were owners of the site, as well as the PCB manufacturer, Monsanto, and the equipment manufacturer, General Electric. The suit was brought to the federal district court for the Eastern District of Pennsylvania. Some plaintiffs brought claims for emotional distress caused by fear of future injury, for medical monitoring designed to decrease the chances of future illness, and for the decrease in value of their property caused by the presence of PCBs on the land. Medical monitoring involves regular physical exams, laboratory tests, and other diagnostic procedures. After holding five days of hearings Judge Kelly entered orders excluding the opinions of all but one of the plaintiffs' experts.[23] This was before the *Daubert* decision by the Supreme Court, but the judge excluded testimony under the *Frye* rule. The appeals court reversed some of Judge Kelly's exclusions but then used the *Daubert* decision, which had recently been ruled, to exclude much of the plaintiffs' expert testimony.

However, the appeals court reversed the summary judgments, so the case went to trial. The trial was divided into two phases, similar to that of the Woburn, Massachusetts, case described in *A Civil Action* and in this book's first chapter. In the first phase, the jury needed to decide whether the plaintiffs had been exposed to PCBs to a greater extent than the general population. The defendants' case for a lack of the plaintiffs' significant exposure was based on the blood test results for PCBs. The Agency for Toxic Substances and Disease Registry of the U.S. Department of Health and Human Services had studied the residents and found that their blood levels of PCBs were no higher than that of the general population.[24] This is another example of measuring the blood level of a persistent chemical to evaluate exposure, similar to the lead level in children described in chapter 7. Lead is persistent in the body because it forms a large reservoir in the bone, whereas PCBs are persistent because it forms a large reservoir in fat. PCBs are fat soluble and resistant to being eliminated from the body because they are not metabolized to make them more water soluble. As a consequence of their ubiquitous presence in the environment at the time of the Paoli litigation, relatively high PCB blood levels were found in the general population. Because the levels measured in the Paoli residents were no higher than in the general population, the jury in this case concluded that there was no evidence that the residents had been significantly exposed to the PCBs in the soil. Since the jury found that they had not been significantly exposed to PCBs from the Paoli railyard, the entire case was dismissed.

The Paoli trial set an important precedent for the determination of whether people who claimed exposure to harmful chemicals needed medical monitoring. The necessary elements for medical monitoring were set forth by the appeals court in an instruction to the jury as follows:

1. Plaintiff was significantly exposed to a proven hazardous substance through the negligent actions of the defendants.
2. As a proximate result of exposure, plaintiff suffers a significantly increased risk of contracting a serious latent disease.
3. The increased risk makes periodic examinations reasonably necessary.
4. Monitoring and testing procedures exist which make the early detection and treatment of the disease possible and beneficial.

Going forward, there has been a growing use of *Daubert* hearings to determine whether expert testimony should be allowed. This trend has added sanity to legal proceedings. What juries hear in the courtroom should be based on good science; otherwise, they have no rational basis for making an informed decision. Unfortunately, in state courts there are often no restrictions on expert testimony, or the less restrictive *Frye* president is used. Plaintiff's attorneys, who decide where suits will be litigated, shop not only states but jurisdictions within states that have histories of allowing unrestrained expert testimony. The result is often huge jury awards that provide the incentive to drive the explosion in litigations that has occurred in the United States.

21

The Toxicology of War

T he uses of toxicology explored previously could be characterized as providing benefit to the public; however, our understanding of the hazardous effects of chemicals can be applied in destructive ways as well, namely, in chemical warfare. We need to understand toxicology's shadow side so that it can be prevented. Importantly, toxicologists need to acknowledge that the underbelly of our field continues to exist and is part of our legacy. For example, the U.S. military conducted controversial toxicological experiments: it exposed its soldiers to chemical weapons. These darkest of experiments were required to determine the dose of chemical warfare agents necessary to incapacitate enemy troops and to determine what protective equipment could shield soldiers against such dosages.

These experiments were the result of the widespread use of chemical weapons during World War I. In July 1915, in a field outside of Ypres, Belgium, the modern age of chemical warfare began when the German army used chlorine gas in a large-scale offensive against the Allies during World War I. The incident rapidly increased attempts on both sides of the conflict toward not only the development of protection against chemical attacks but also the production of more effective chemical weapons. Chlorine and phosgene gases were the first chemical weapons manufactured, but after a properly designed gas mask was invented to protect soldiers' lungs, they were no longer effective.

Subsequently, the chemical called mustard gas was developed, and its vesicant properties, causing blisters on the skin, eyes, and in the lungs, were used to immobilize the enemy. In July 1917, mustard gas was used by the Germans for the first time.[1] It was not a systemic poison and didn't need to be absorbed into the bloodstream to have an effect. Because of this, it acted almost immediately; it also was persistent, which was an advantage in the battlefield. Mustard gas caused almost four hundred thousand casualties during the war, many more than any other chemical agent.[2] It was estimated that there were ninety thousand deaths and one million casualties caused by all chemical weapons used during World War I. The U.S. military sustained approximately seventy thousand of these casualties.[3]

The United States began its development of chemical weapons in 1917 after joining World War I, and it developed two laboratories at the newly formed American University and at Catholic University of America in Washington, DC. Winfred Lee Lewis was in charge of the effort, and the next agent developed was called "lewisite."[4] In the United Kingdom, research and development of chemical agents began earlier in 1916, at Porton Down.

Between the two world wars, a number of countries further developed their chemical warfare capabilities. Germany, Italy, Japan, Britain, France, the Soviet Union, and the United States were stockpiling chemical weapons. Japan did not use chemical warfare during World War I, but by the mid-1930s Japan was manufacturing enormous quantities of poison gas bombs, including shells of chlorine, phosgene, and mustard gas. In October 1935, Italian forces crossed into the territory of present-day Ethiopia and successfully deployed chemical weapons, primarily mustard gas. An estimated 15,000 of the fifty thousand Ethiopian casualties in the war were caused by chemical weapons.[5]

• • •

The U.S. military felt compelled to test protective equipment to shield soldiers from the possible use of mustard gas. A testing program was initiated by the War Department in order to study methods of protecting soldiers who might be exposed to these agents. The military testing program during World War II involved about sixty thousand subjects. Patches of mustard gas or drops of the chemical were the most commonly used methods to assess the ability of ointments to protect or decontaminate the skin. In chamber tests, soldiers were put in small rooms and exposed to mustard gas in order to assess the dose response. These chambers were also used to gauge the ability of masks to protect the respiratory tract and test specially treated clothing intended to protect the skin. In some cases, the men were exposed repeatedly; they entered the chambers either every day or every other day until they developed moderate to intense reddening of their skin. Chamber and patch testing in the United States was done at Edgewood Arsenal, Bainbridge, Maryland; Camp Sibert, Alabama; the Naval Research Laboratory, Virginia; Camp Lejeune, North Carolina; and San Jose Island, in the Panama Canal Zone. According to the Institute of Medicine, one thousand U.S. servicemen also participated in large-scale field tests to determine whether protective clothing would work under battlefield conditions.[6]

Dose-response information for each of the various exposure pathways was documented for mild up to incapacitating effects either from these chamber experiments or from studies available from battlefield exposures. Heat and humidity were found to increase the effectiveness of mustard gas on the skin; severe effects were ten times more likely on sweating skin than dry skin. The eye was more sensitive than the skin, probably because of the moisture from the tear gland, resulting in corneal damage. Acute lung effects could lead to death from bacterial infections such as pneumonia.[7]

Various protective agents were tested, including masks, clothing, skin creams, and gloves. The chamber tests sought to answer several important questions. How long would the protective measures be effective for? What did the environmental conditions do to exposure? What concentrations of gas would be neutralized by clothing impregnated with carbon or some other chemical? There were instances where some permanent damage was found to occur in spite of the use of protective equipment.[8]

Gas was not used by the Germans during World War II. The reasons for this are still debated, but one reason certainly was fear of retaliation. One story describes Hitler as personally opposed to using gas because he was exposed to it by his own army during World War I. The wind direction could change during deployment of these weapons, which made them tricky to use. On the other side, some Allied soldiers had been exposed in World War I and too were reluctant to unleash the gas genie. As a result, the U.S. policy was not to use poison gas unless the enemy did.[9]

• • •

In World War II, the Germans stockpiled not only mustard gas but also paralyzing nerve agents. The organophosphorus nerve agents, including sarin, soman, and tabun, were synthesized by German scientists in the 1930s and 1940s and were originally intended to be used as agricultural insecticides. The Allies did not know of their existence and had no defense against them, so it was fortunate that Germany did not use them. At the end of World War II, the Russians moved a captured manufacturing facility for these weapons (and its personnel) to the Soviet Union and continued production. Following World War II and the discovery of these events, the U.S. and UK military initiated large research programs to develop these compounds for military applications.[10]

Because organophosphorus nerve agents are essentially colorless, odorless, tasteless, and nonirritating to the skin, their entry into the body may not be perceived by the victim until grave signs and symptoms appear. When symptoms did arise, incapacitation or death soon followed, given their extremely high acute toxicity.[11] These nerve agents are chemically similar to organophosphorus pesticides such as chlorpyrifos (Dursban, Lorsban), parathion, malathion, and acephate (Orthene), which have been widely used in agriculture.[12] Agent VX was later developed and is by any route of exposure the most potent of all the nerve agents.[13]

The organophosphorus nerve agents exert their pharmacological effects on the nervous system through inhibition of the enzyme acetylcholinesterase. During nerve transmission, the neurotransmitter acetylcholine is released from the presynaptic nerve into the synaptic cleft. Receptors on the postsynaptic neural membrane bind acetylcholine, thereby relaying the signal to the next nerve cell or muscle. As described in chapter 2, inhibition of this receptor is the mechanism by which the sea snake's venom has its effect: it paralyzes the snake's prey. Acetylcholinesterase is responsible for breaking down the acetylcholine. Chemicals that inhibit or inactivate this enzyme allow the neurotransmitter acetylcholine to accumulate in the synaptic cleft, effectively leading to a state of continuous nerve stimulation. Eventually this continuous stimulation deadens the ability of the nerve or muscle receptor to respond, causing respiratory paralysis and death.[14]

In the mustard gas experiments, the effects were primarily studied in humans because the gas's toxic effects were specific to the human skin, eye, and respiratory tract. However, with nerve agents, experimental animals were primarily used because the mechanism of action on acetylcholinesterase was similar to that in humans. So in the United States, research on nerve agents involved studies in animal models as well as some use of both military personnel and civilian subjects. To provide some level of protection for human subjects, the agents were administered by several routes in at least seven species of animals before being used in human testing.[15] Similar testing programs were conducted in the United Kingdom at Porton Down; these continued well past World War II.[16]

After acetylcholinesterase inhibitors were developed as agents of chemical warfare, part of the continued military testing program conducted in the 1960s and early 1970s involved finding antidotes for nerve gas poisoning. Drugs such as atropine antagonize some of the effects of the nerve gases, but they fail to work on other effects. Acetylcholinesterase reactivators, which reverse the effects of nerve gases by making the receptors function again, were another treatment.

Eventually, the United States terminated its program of testing and production of chemical weapons. President Nixon ordered the limitation of chemical-biological warfare production in 1969 and the destruction of stockpiles of certain chemical weapons. In late 1974, the U.S. Senate voted to adopt the 1925 Geneva protocol on chemical warfare.[17] The Department of Defense Authorization Act of 1986 directed and authorized the secretary of defense to destroy the United States' stockpile of lethal chemical munitions and agents by September 30, 1994.[18]

•   •   •

Mustard gas has been used in other, more recent conflicts. In 1963–1967, it was used by Egypt's President Nasser against Yemen.[19] In the Iran-Iraq war of 1980–1988, mustard gas and nerve agents were again used in large amounts by the Iraqi leader Saddam Hussein against Iranian forces and his own Kurds. The Iraqis used mustard and nerve gases against Iranian offensives in southern Iraq. In March 1988, Hussein killed between 3,200 and five thousand Kurds around the town of Halabja and injured thousands more, most of them civilians.[20]

Even though much was learned about nerve agents in the experimental tests on experimental animals and soldiers, two exposure episodes by terrorist attacks in Japan using sarin gas provide the best documentation of the effects of high exposures to nerve agents. In the late evening of June 27, 1994, Japanese terrorists spread sarin vapor, using a heater and fan mounted on a truck, in a residential neighborhood near the center of Matsumoto, Japan. About six hundred people (residents and rescue teams) developed acute symptoms of sarin exposure (acute cholinergic syndrome); fifty-eight people were admitted to hospitals, 253 sought medical assistance, and seven died.[21]

On the morning of March 20, 1995, a religiously motivated cult released sarin in a terrorist attack on five subway cars on three separate subway lines. The attack was timed to coincide with peak commuter traffic and targeted at a subway convergence point underneath the Japanese national government's ministry offices. Eleven commuters were killed, and more than five thousand people required emergency medical evaluation. Victims of the attacks were treated with atropine and acetylcholinesterase reactivators.[22]

Regarding chemical weapons, we have let the toxic genie out of the bottle. The elimination of the stockpiles of these weapons by the United States and other countries can go a long way toward ending this global threat. Hopefully we will never again have to use soldiers in chemical weapon experiments, even for the purpose of protecting large numbers of soldiers. Research and improvements in animal testing for these and new agents can lessen the risks for soldiers. The recent destruction of Syria's chemical weapon stockpiles was thought to eliminate the last major potential source of these poisons. But we shall see. There have been continuing signs that such weapons are being used in Syria despite efforts to control their use.[23]

IV | The Unfinished Business of Toxicology

Franklin. *Eh! Oh! Eh! What have I done to merit these cruel sufferings?*
Gout. *Many things: you have ate and drunk too freely*
And too much indulged those legs of yours in their indolence
Franklin. *Who is it that accuses me?*
Gout. *It is I, even I, the Gout.*

—Benjamin Franklin, *Poor Richard's Almanac*

This last section describes the most interesting challenges in toxicology today and places them in their historical contexts. First we examine the history of opiate addiction, which has led to the contemporary overdose crisis. This topic is given little attention in both toxicology and occupational medicine unless the topic of drug testing is discussed. A wider and more in-depth discussion is deserved, with more emphasis on science, as the brute law enforcement approach has clearly not worked.

Next we will cover the toxic effects of air pollution, which cannot be addressed without discussing its incredibly important link to climate change. Toxicologists are becoming increasingly aware that the health and climate problems caused by the production and burning of fossil fuels are critically entwined. The immediate and obvious problem of health-threatening air pollution from the use of fossil fuels could be a more translatable and urgent catalyst for change than the long-term and, arguably, somewhat obscure effects of climate change.

The next group of chapters addresses the complexities and problems associated with using animal models for studying and treating human disease. This subject has been a common thread throughout this book. In chapter 24, we explore the more general issue of animal models as they relate to the study of human diseases, including cancer development and chemotherapeutic agents. The translation of this intraspecies research has been disappointing in many instances; however, the relative ease of animal research has prompted its continued use. In chapters 25 and 26, we examine tests of chemicals in animal bioassays and *in vitro* tests of cells to predict human cancer and hormonal effects. The interpretation of these results for their application to humans requires extensive additional mechanistic research, which in most instances is lacking. Moreover, many governmental regulatory agencies do not even recognize this need. There is some promise in proposed alternatives to the conventional rodent bioassays for testing chemicals, as we will discuss in chapter 27; however, these alternatives need to be verified for application to human disease. In the final chapter, we look to future directions for disease prevention, with further implementations of toxicology to some unresolved public health problems. Our studies of the toxic effects of chemicals have made tremendous progress in our understanding of disease and strategies for its prevention; unfortunately, the personal and political barriers to implement this knowledge are often formidable.

22 | Opiates and Politics

Drug overdoses and addictions are the two aspects of arguably the most serious and intractable public health crisis we face. More people died from drug overdoses in the United States in 2014 than during any previous year on record. From 2000 to 2014 nearly half a million people in the United States died from drug overdoses. In 2014 there were approximately 1.5 times more drug overdose deaths in the United States than deaths from motor vehicle crashes. In 2014, 61 percent of drug overdose deaths involved some type of opioid, including heroin.[1] How did we get into this mess, and what can toxicology do to help? Over the next three years, the rates continued to increase, to 72,000 deaths in 2017, double the rate compared to the average of that for 2000 to 2014. The recent increases were caused by synthetic opioids such as fentanyl.[2] By comparison, there were fewer than three thousand overdose deaths in 1970, when a heroin epidemic was said to be raging in U.S. cities and the "War on Drugs" was pronounced by President Nixon, and there were fewer than five thousand recorded in 1988, at the height of the crack epidemic.[3]

Perhaps no other problem in toxicology has as much of a political component as drug abuse policies. The problem with politics is that they are often toxic, and the mix of opiates and politics is particularly so. Add to that the stereotype of the lives of crime and vice associated with many of the victims of overdose and addiction, and we now have a complicated mix of toxicology, sociology, mental health, law enforcement, and politics. But first, let's explore a bit of opiate history and research, elucidating how these chemicals cause their toxic effects.

The dried milky fluid from the poppy plant was first described in the Ebers Papyrus, the first record of ancient Egyptian medicine. Written in about 1500 BCE and describing practices from one or more millennia earlier, the Ebers Papyrus described poppy as useful for headaches and as anesthesia. The Ebers Papyrus included the description of a "remedy to prevent excessive crying of children." Opium had also been used to make people happy since the times of ancient Greece. In Greek, *opion* means poppy juice, and in Homer's Odyssey, "A new thought came to Zeus-born Helen; into the bowl that their wine was drawn from she threw a drug that dispelled all grief and anger and banished remembrance of every trouble. . . . Presently she cast a drug into the wine of which they drank to lull all pain and anger and bring forgetfulness of every sorrow."[4] And it was Paracelsus who, having experimented with various opium formulations, discovered that the alkaloids in opium are far more soluble in alcohol than water and came across a specific tincture of opium that was considerably useful in reducing pain. He called this preparation *laudanum*, derived from the Latin verb *laudare*, to praise.[5]

Opium is produced from the newly formed seed pod of *Papaver somniferum*. The pod is cut, allowing the juice to seep out, and the latex is then collected after drying. In 1803 a German pharmacist, F. W. Sertürner, isolated the active ingredient from opium and named it "morphine," after Morpheus, the Greek god of dreams. Morphine was found to be about ten times more potent than raw opium.[6]

In the United States, the increase in opium and morphine use is usually attributed to treatment of soldiers during the Civil War, even though morphine injection then was relatively rare. Nearly 10 million opium pills were issued to Union forces. The growing availability of syringes after the war allowed morphine to be injected, leading to an increase in addiction from

1865 to 1895 among veterans. Also during this period, patented medicines that contained opium and morphine became increasingly available.[7]

C. R. Alder Wright in 1874 experimented with morphine derivatives and produced diacetylmorphine, also known as heroin. However, not much was made of this discovery until the German pharmacologist Heinrich Dreser, of Friedrich Bayer and Co., reported its synthesis in 1898 at the seventieth Congress of German Naturalists and Physicians. The drug was initially used in powder form and taken orally for respiratory diseases such as pneumonia and tuberculosis. It was found to reduce coughs dramatically and produced a strong sedative effect on labored respiration. Heroin was also reported erroneously to have a much greater therapeutic index than morphine, with a lethal dose a hundredfold greater than its therapeutic dose. It was also claimed that heroin could be used as a treatment for morphine addiction; of course, the result was heroin addiction. The Bayer Company advertised the drug in a number of languages along with another famous compound from their laboratory: aspirin.[8]

Not only were opiates used as a medicine, but they also gained increasing appeal for their mind-altering properties. Recreational smoking of opium in the United States was initially confined during 1850 to 1870 primarily to Chinese immigrants who had come from a long tradition of opium smoking. After 1870 opium addiction began among the non-Chinese population, especially prostitutes, gamblers, and petty criminals. But the attraction of opium smoking was eventually felt by the upper class and particularly the "idle rich" as well. The publication in 1821 of Thomas De Quincey's *Confessions of an English Opium Eater* may have had some influence on the use of opiates among intellectuals, but it is not clear how much this contributed to addiction in the United States.[9]

• • •

Legal efforts to check opium smoking began in 1880, when New Hampshire entertained a bill to increase the duty on imported opium. In 1909, the United States banned all importation and possession of opium, which had the unintended consequence of forcing addicts to obtain cheaper morphine and heroin.[10] Consequently, the recreational use of heroin increased at the beginning of the twentieth century, and its injection followed soon after. Injection vastly increased the euphoria and the possibility of death from overdose. In 1917 Charles Stokes, at Bellevue Hospital in New York, reported that ten of the eighteen heroin addicts evaluated used hypodermic syringes to administer the drug.[11]

The Harrison Narcotic Act of 1914 was intended to tax opium and coca leaves and their derivatives. Registered physicians were required under the law to keep records of the drugs that they prescribed or dispensed. According to the Consumer Union Report *Licit & Illicit Drugs*, "It is unlikely that a single legislator realized in 1914 that the law Congress was passing would later be deemed a prohibition law. The provision protecting physicians contained a joker—hidden in the phrase, 'in the course of his professional practice only.'" Law enforcement interpreted this clause to mean that opiates could be prescribed for pain but not for the maintenance of an opiate addiction. Opiate addiction was not considered by the law to be a medical condition. This view led to the prosecution and imprisonment of physicians who tried to treat their addicted patients with opiates.[12]

The development of the concept of addiction as a disease requiring treatment had its beginning in the United States in 1919 as a result of the tightening opiate supply after the Harrison Narcotic Act. So-called narcotics clinics were developed to treat opiate addicts by maintaining them on heroin, morphine, or opium while providing treatment for their other medical problems. Many of these clinics were extensions of facilities to treat tuberculosis, venereal disease, and mental illnesses. The need was great, and within three months of opening, three thousand addicts appeared on the doorstep of the Worth Street clinic in New York City. Unfortunately, this and most of the other clinics were quickly closed by the Treasury Department, which believed that addicts were criminals, not medical patients.[13]

One of the most successful clinics was that of Dr. Willis Butler, at the Schumpert Memorial Sanitarium, the largest hospital in Shreveport, Louisiana. His approach to the treatment of opiate addicts was first to prevent withdrawal and then attend to their medical problems. When the organic illnesses were under control, attention was next focused on the addiction. Patients deemed capable of resuming their normal lives had their intravenous dosages slowly decreased and were often treated with oral opiates or sedatives to ease the withdrawal. Finally, once a patient was drug free, they were kept in the hospital for up to a month for observation. Some patients who did not have medical problems were detoxed from the beginning of their hospital stay. Treasury Department agents harassed Dr. Butler to the point that his last narcotics clinic was closed in 1923, even though it was considered a success.[14]

Opioid maintenance therapy was not to be seen again for over forty years in the United States. In 1964 the doctors Vincent Dole and Marie Nyswander began medically treating heroin addicts with methadone, a synthetic opiate that lasted for about twenty-four hours, at an experimental addiction treatment program at the Rockefeller Institute in New York City. Vincent Dole's background was in metabolic diseases, and he likened the effects of methadone to that of insulin for diabetics and cortisone for arthritics. Nyswander, a psychiatrist, along with Dole proposed that heroin addicts were organically ill and were not necessarily psychologically disturbed just because they were addicts.

In 1968 Dole and Nyswander published their report of their four years of experience with 750 criminal addicts in the *Journal of the American Medical Association.* They observed that when addicts were gradually stabilized on substantial doses of methadone, they did not experience opiate withdrawal and did not crave heroin. Also, they did not get high from injected heroin. Thus the theory of the methadone blockade was born. Methadone was an opiate agonist, but as the dose was gradually raised, opioid tolerance developed. Higher doses of methadone also acted as an "antagonist" against the effects of other opioids, including injected heroin; therefore, it prevented overdoses from heroin. Another convenience of methadone, as opposed to morphine maintenance, was that it could be given orally on a daily regimen.[15]

The real revolution of methadone maintenance was its emphasis on medical, social, and psychological rehabilitation rather than merely on heroin detoxification, which was the only method of treatment at that time. This reversed the approach to addiction of the previous four decades, which had assumed that abstinence must come first and that rehabilitation was impossible while a person was taking drugs of any kind. The theory behind methadone maintenance was similar but more expansive than that of Butler's heroin clinic in Shreveport forty years earlier. Butler believed that after a patient was stabilized on heroin, they could achieve a stable way of life, with a job, a home, a position of respect in their community, and a sense of worth. Then it would be possible to discontinue heroin and achieve a drug-free state. In contrast, Dole and Nyswander proposed long-term, often lifelong maintenance on methadone, similar to a diabetic's mainte-nance on insulin. As a result, since the 1970s methadone maintenance, rather than detoxification, has become the main medical treatment for opiate addiction.[16]

$$\bullet \bullet \bullet$$

Richard Nixon declared his war on drugs on June 17, 1971, by setting up the Special Action Office for Drug Abuse Prevention (SAODAP) in the Execu-tive Office of the President. Before this, federal drug abuse research was sub-sumed under the NIMH, in the Center for Studies of Narcotic and Drug Abuse (CSNDA), which was staffed by about ten professionals, and clinical research was done at the Addiction Research Center in Lexington, Ken-tucky. The SAODAP was established in response to an expansion in heroin addiction, including in soldiers returning from Vietnam; Jerome Jaffe, and a psychiatrist and drug researcher from the University of Chicago, was put in charge. However, the White House kept close control over this high-priority activity. Egil (Bud) Krogh, who would later become infamous for his role in the "plumbers' unit" during the Watergate scandal, personally supervised all SAODAP activities.

The aim of the SAODAP was to coordinate for five years all activities involving drug abuse prevention, research, and treatment in all governmen-tal agencies and departments. Jaffe needed to staff up the SAODAP quickly, which would include hiring over one hundred personnel. Because the NIH had doctors that could be tapped, Jaffe found Alan Greene, who was a com-missioned officer in the Public Health Service at the National Institute of Mental Health, and made him his assistant.[17] Two years later, in April 1973, I would also leave my Public Health Service post at the NIH and replace Alan, who was then coordinating the biomedical research efforts for the office. By this time, methadone maintenance had become the gold standard for heroin addiction treatment. The federal government supported an expansion of community treatment programs and Veterans Administration programs to provide methadone and other drug treatment across the country. In addition, in 1973 the National Institute on Drug Abuse (NIDA) was created to continue the efforts of the SAODAP and give more permanent emphasis and prominence to drug abuse research and treatment.

The SAODAP's budget for research, treatment, and prevention of opiate addiction went from $146.5 million in 1971 to $446.8 million in 1975.[18] That would be over two billion in today's dollars (NIDA's budget in 2017 was about one billion dollars). In comparison, in 1973 the newly formed Drug Enforcement Administration (DEA) began its work with an annual budget of $75 million; by 2014, the DEA's budget was approximately $2 billion.[19]

What kinds of research were the SAODAP and NIDA funding? The discovery of opiate receptors by Solomon Snyder, at Johns Hopkins Medical School, was difficult—certainly not as easy as experiments with the bungarotoxin sea snake venom and the acetylcholine receptor had been. The difference was the degree of opiate binding and specificity. Whereas the snake venom binds specifically and almost irreversibly to the receptor, opiate receptors are relatively scarce in the brain, and opiates bind nonspecifically to many sites throughout the body. The breakthrough for Snyder came when using the radiolabeled antagonist naloxone, which bound to the one type of opiate receptor more strongly than the opiates did.[20]

Opiate receptors not only have to do with attenuating the perception of pain but also with other functions throughout the body. One of these has to do with the autonomic nervous system's respiratory control system. If there is too much opiate receptor stimulation, breathing is depressed. As opposed to snake venoms, which inhibit the neuromuscular junction, opiates affect the central nervous system's regulation of breathing. But the ultimate effect of too much opiate is the same: death from asphyxiation.

The next step was to determine why the opiate receptor is normally present in the brain, and some of the research involved identifying chemicals produced by the brain to bind with the opiate receptor. These endogenous substances were difficult to find because unlike the venom in a snake, which is highly concentrated, these endogenous substances were estimated to represent only a tiny fraction of the weight of the brain. Researchers had to process huge numbers of brains, purifying substances and testing them to determine whether they had opiate effects. Eventually two types of five-amino-acid peptides, which are like small proteins, were identified by Hans Kosterlitz and John Hughes, from the University of Aberdeen, in Scotland, along with the chemist Howard Morris, from the University Chemical Laboratory, Lensfield Road, Cambridge. These peptides were called enkephalins; they bound to the opiate receptors. Eventually other larger polypeptides would be identified, endorphins, that bound to other opiate receptors and had similar and diverse functions throughout the body, like their exogenous equivalents opium, morphine, and heroin.[21]

• • •

A major problem with methadone was that it lasts only about twenty-four hours; consequently, either the addict has to come to the treatment clinic daily for supervised dosing or be given take-home doses. Take-home dosing created an illicit trade in methadone, and the newly formed DEA and local police were applying pressure to close the methadone clinics. Jaffe, along with other researchers, had been studying a longer-acting opiate called levo-alpha-acetylmethadol (LAAM), which only needed to be taken three times per week. This increased the practicality of limiting take-home doses and the subsequent diversion of methadone from the treatment clinics to the street.

Besides the development of chemotherapeutic agents by the National Cancer Institute, LAAM was one of the first, if not the first, "orphan" drug that another institute, in this case the National Institute on Drug Abuse, wanted to develop. They had hoped to get a major pharmaceutical firm to perform the final phase of work, which included studies of three thousand patients and the filing of a New Drug Application with the FDA. They sweetened the deal to include the exclusive rights to market the drug. Unfortunately, the small projected market of heroin addicts made it so unattractive that no established pharmaceutical company wanted to take on the research and subsequent sale of the drug. There was only one bidder besides my company. By this time, I had left the government and was performing government contract work on lead-based-paint poisoning prevention. So I decided to bid on the LAAM contract. My only competition was not a drug company but another consulting firm that apparently had neither the expertise nor an adequate plan to do the study.[22]

I was awarded the contract, but the major problem had nothing to do with the conduct of the study: the problem became Washington politics. Jack Anderson was the preeminent muckraking journalist of the time and was published nationally in many newspapers, including the *Washington Post*. He was systematically attacking the Carter administration, which was then supervising my contract. Carter's White House press secretary Jody Powell charged that Anderson had become "recklessly irresponsible" and undertaken a "vendetta" against the Carter administration.[23]

Anderson investigated the National Institute on Drug Abuse, and one of his articles, published on July 1, 1978, was about the LAAM study. Most of this article accused some clinics in Los Angeles of coercing patients into the study and not obtaining adequate informed consent. I had followed up on this with the clinics and was assured that proper procedures were being followed, recognizing that there is a degree of subtle coercion required to keep heroin addicts in any type of treatment program. Almost as an aside and in the last part of his article, Anderson implied that I had received my contract improperly and was receiving a gift of the drug at the taxpayer's expense.[24]

Nothing much came from this article, and though hearings were held by Representative John Moss, the matter was dropped. But more trouble would come the following year when Representative Henry Waxman became chairman of the Subcommittee on Health and the Environment. His staff got together with Howie Kurtz (who had been Anderson's reporter), held hearings on my LAAM contract on March 27, 1979, and demanded to know from NIDA Director William Pollin why my contract gave me the rights to the drug in return for cost sharing and my pledge to market and distribute the drug. Pollin answered Waxman's question and defended NIDA's actions. The day following the hearings Kurtz wrote the first of a series of articles published in the now-defunct *Washington Star* entitled "U.S. Drug Abuse Agency Draws Health Panel's Wrath." Eventually, the political heat became too much, and even though the project manager on our contract, Jack Blaine, and his boss, Pierre Renault, were supportive, Secretary of Health, Education, and Welfare Joseph Califano bowed to Waxman's demands and decided that a modification of the contract to include women be halted.

By that time I had already submitted the NDA application for LAAM's use in men, involving thousands of pages of reports and hundreds of boxes of patient data. The *Washington Post* wrote an article, "Federal Contract: A Litany of Frivolity, Waste," that vindicated my predicament. My project officer, Jack Blaine, was quoted, "It was the government bureaucracy that botched it up. It was politics. The thing is, the contractor did a very good job." However, even though the government believed that they could continue without my contract, the development of the drug languished after that.[25] Several years later NIDA contracted with a colleague of mine, Alex Bradford, to complete the rest of the study. The drug was marketed and then used successfully in clinics after 1993, when the New Drug Application was approved.

Later, reports of the potential for life-threatening ventricular arrhythmias seen in an EKG abnormality raised concerns about potential cardiac effects both of LAAM and methadone. In the context of these cardiac side effects, the European Medicines Agency (EMA) recommended the suspension of the marketing authorization for LAAM in 2001, and the U.S. Food and Drug Administration (FDA) required the addition of a "black box" warning on the LAAM label.[26] Studies showing the higher efficiency of LAAM with respect to suppression of heroin use then revived the discussion of whether potential risks for cardiac arrhythmias really outweighed the advantages of LAAM.[27] The problem is that the government would need to screen people with EKGs, and this would cost money.[28]

• • •

The number of patients receiving methadone at the beginning of 1977 was estimated to be ninety thousand, with another sixty thousand in largely narcotic-free rehabilitation programs. Many of these patients were veterans being treated in VA clinics.[29] In comparison to 1977, the number of methadone patients had increased from ninety thousand to only 117,000 in 1993, which may have reflected a lull in the heroin epidemic following the initial efforts of the SAODAP and NIDA.[30] However, in 2003, the number had risen to 227,000 and in 2015 to 357,000; this was from the explosion in the rate of addiction by other opiates.[31] Methadone maintenance remains an effective treatment option, but its use is limited because it is only available at specialized methadone treatment clinics.

So let's look at other options for treating opiate addiction. Another synthetic opioid, buprenorphine, is only a partial opioid agonist, which means it eliminates opioid craving and withdrawal while blocking the effects of heroin and other opioid analgesics. Buprenorphine was approved as a Schedule III controlled substance, which puts it in the same category as Tylenol with codeine, meaning that it can be prescribed by doctors in their regular practice, although under restrictions. Consequently, it may be the most available opioid addiction maintenance treatment, which is fortunate given the recent opioid epidemic.[32]

Naltrexone is a different type of maintenance: opiate antagonist maintenance. This drug can be viewed as a long-acting form of naloxone (Narcan) that blocks the opiate receptor and has no or minimal agonistic effect. Therefore it has to be administered after the patient is opiate drug free. It was eventually approved through the efforts of NIDA and Alex Bradford, the investigator who finished the development of LAAM. Initially the oral drug did not do well, because an addict could stop taking it and get high on heroin after a day or two. It is now making a comeback as the intramuscular injection Vivitrol, which lasts for a week.

Events in Portugal after 2001 provided yet another solution to the opiate addiction crisis and a blueprint for how to prevent the problems caused by "criminal" addicts. In that year, the Portuguese parliament decriminalized the use and simple possession of all drugs; consequently, addicts were no longer considered criminals. However, it was still illegal to sell drugs; that is, Portugal did not renounce the UN conventions authored by Harry Anslinger, the first commissioner of the U.S. Treasury Department's Federal Bureau of Narcotics. Following the decriminalization of drug use, predictions of catastrophe were widespread. It was thought by Portuguese law enforcement officials, such as the chief of the Lisbon Drugs Squad, Joao Figueira, that the use of drugs would explode. Instead, Figueira later conceded that "the things we were afraid of didn't happen."[33]

What actually happened not only astonished the law enforcement officials but also surprised supporters of the new law. By being able to evaluate drug use without legal constraints, they found that 90 percent of drug users did not have a serious problem and should be left alone, allowing resources to be spent on the 10 percent of addicts who needed help. To provide adequate help for these addicts, the machinery of the drug war was turned to treatment programs. Previous spending for law enforcement activities aimed at drug use and possession was instead used for treatment of addiction.[34]

What was the result of this radical approach? First, the lives of addicts transformed, and they were no longer robbing people for their next fix. The number of problem drug users decreased, injecting drug use was cut in half, the number of overdoses was reduced significantly, and the proportion of people contracting HIV was reduced to less than half of its original rate. Crimes on the street related to drug use virtually disappeared because addicts were either in treatment on methadone or recovering from addiction. To help matters even more, the government gives a large, year-long tax break to anyone who employs a former addict.[35]

So there are a number of possible new solutions to the opiate epidemic and overdose problem. However, there are also new challenges that are making matters worse. In 1991, in one of the worst cases of multiple overdose in the New York area, seventeen people died from fentanyl, a powerful synthetic drug that was being sold as heroin. Fentanyl was used for decades to sedate patients undergoing surgery. The drug is one hundred times more powerful than heroin, cheaper to make in the laboratory than heroin, and easier to smuggle into the United States.[36] More recently fentanyl has become a major cause for alarm. Between 2013 and 2014, rates of death involving methadone remained unchanged; however, death involving natural and semisynthetic opioid pain relievers, heroin, and synthetic opioids other than methadone (for example, fentanyl) increased 9 percent, 26 percent, and 80 percent, respectively.[37]

To summarize, opiate addiction and overdose remain two of the most vexing problems in society. We have learned that drug abuse treatment is based on the premise that people who are addicted have problems related to medical disorders, psychiatric illnesses, or social conditions. Drug addiction can lead to death or debilities from overdose or withdrawal. Therefore, it is important to get addicts into treatment programs and provide adequate therapy for all the underlying medical conditions, among others, that led to drug abuse in the first place. Then again, the law enforcement side of government treats addicts as criminals and also views doctors who treat addicts with suspicion. The solution of this problem needs to involve out-of-the-box thinking like the approach used in Portugal.

23 | The Toxicology of Climate Change

E very day in the pediatric emergency room at Bronx Municipal Hospital in 1970, where I was a pediatric intern, the halls
were lined with children; most of the kids were in the hospital because they had asthma. In some ways, the asthmatics were
the easiest patients because they were familiar with the routine
and provided little resistance to examination and treatment.
These children were struggling to breathe and knew that relief
was just a shot of epinephrine away. Even though immediate
treatment was possible and effective, most of these young asthmatics would come back, some even later that same day.

Why were these children sick? Asthma can be a complicated
disease; however, one significant risk factor is air pollution, which
in the Bronx would have come primarily from the burning of
truck fuel and sulfur-containing oil for heat. We now know that
when sulfur dioxide and water vapor mix, sulfuric acid is produced, causing constriction of the small airways in the lungs.
This is exacerbated by a tendency of the sulfuric acid to get
trapped in the liquid coating of the upper airways. These mechanisms ultimately combined forces to make breathing, especially
exhaling, difficult. This was first shown experimentally by Mary

Amdur, at New York University in 1952, by exposing men to varying concentrations of sulfur dioxide via face mask. Later, at the University of Rochester Mark Utell and his associates exposed patients in chambers and found that asthmatics were about ten times more sensitive to sulfuric acid aerosols than normal adults, with significantly impaired breathing at the levels of sulfuric acid common in the urban atmosphere.[1] This study was confirmed by Jane Koenig and her associates at the University of Washington, using sulfur dioxide administered via mouthpiece.[2]

The health effects from air pollution containing sulfur dioxide were not new phenomena. The use of sulfur-containing coal for heat in households led to the notable pollution described by many writers, including Charles Dickens, who lived in London during the nineteenth century.[3] During some winters of the 1950s there were long periods of dense fog in London, and visibility was reduced to under two hundred yards.[4] Medical staff at St. Bartholomew's Hospital reported that this fog contained increased concentrations of both smoke and sulfur dioxide. In December 1952, the total number of deaths from the fog was estimated to be about four thousand in the Greater London area.[5] The number of deaths caused by the fog in the winter of 1954–1955 was reported to be about one thousand. Deaths, mostly attributed to bronchitis, disproportionately affected newborns and the elderly. In those days, statistics did not differentiate between asthma and what we now call bronchitis, but with the benefit of hindsight we can say that much of this harm was probably asthma related and exacerbated by sulfur dioxide.

In addition to sulfur dioxide, coal combustion produces polyaromatic hydrocarbons (PAHs) and emits metals such as nickel, chromium, arsenic, and mercury. These PAHs are the same cancer-causing chemicals found in the chimney soot Percival Potts identified as the cause of scrotal cancer, and they are found in cigarette smoke as well. The elements nickel, chromium, and arsenic are also known to be human carcinogens. These carcinogens are bound to larger particulates, and adequate treatment of the chimney exhaust in power plants can reduce their release to acceptable levels. Two techniques are used to capture these particles. Filtration of the air can be done in "bag houses," which have something like a large vacuum cleaner bag through which the exhaust must pass. In electrostatic precipitation, the gas flows through electrodes that induce negative charges on the particles, which then become attached to positively charged collectors.

Such controls do not work very well for mercury emissions because mercury is a gas at smokestack temperatures. Thus mercury poses a challenge for controlling pollution from coal burning and requires expensive additional controls. The reduction of mercury releases from coal plants has been a major focus of public health initiatives because of mercury's effects on the nervous system and the kidneys.[6] Releases of mercury into the atmosphere can be transformed into methylmercury in fresh water, and this much more toxic form of mercury can reach humans through fish consumption. The nervous systems of unborn babies and children are most susceptible to its effects.[7]

And it's not just air pollution from coal production that needs to be considered. The hazards of coal mining have been described for centuries. René Laennec (1781–1826), a French physician best known for inventing the stethoscope, found that inhalation of coal dust caused a sometimes fatal disease that he called melanosis, known colloquially as black lung disease.[8] The yearly death rate from diseases of coal miners has been estimated to be twenty-five to thirty thousand workers per year worldwide from 1990 to 2013.[9]

• • •

Analysis of the climate impacts of energy production necessarily becomes political, positive changes come very slowly, and it is difficult for the scientific evidence to make a difference because of the abstract nature of future predictions. On the other hand, toxicologists recognize that the production and burning of fossil fuels, which also contribute to climate change, have serious and immediate detrimental effects on an individual level, such as the asthmatics in the pediatric ER. Consequently, toxicology provides the key insight that climate change and asthma caused by air pollution, two seemingly unrelated issues, share a single direct cause. While the impact of fossil fuels might be hard to conceptualize in terms of faraway shrinking polar ice caps or the gradual drowning of coastal cities, it is easy to understand the disease burden that these activities place on the most vulnerable members of society. As we shall see, toxicology shows us that the incremental causes and effects of climate change have immediate implications in terms of both occupational and environmental health.

For those in need of a refresher on the basics of climate change, it works roughly as follows. The combustion of fossil fuels produces other chemicals in addition to the asthma-producing sulfur dioxide, the most prevalent being carbon dioxide, which accumulates in the atmosphere. When sunlight penetrates the atmosphere, it generates heat. Typically, a portion of this heat warms Earth's surface, and the remainder is released back out through the atmosphere. However, accumulated carbon dioxide in the atmosphere from burning fossil fuel prevents some of this heat from escaping. This process is called the greenhouse effect and is responsible for much of global warming.

The consequences of the rise in temperature from the greenhouse effect are of great international concern. To quote the Climate and Health Program at the Columbia School of Public Heath, "Climate change affects health through complex mechanisms that include shifts in the global atmosphere, in regional ecology, in social structures, and in human exposures and behaviors." For example, one probable result of climate change is the future spread of mosquitoes that transmit horrible diseases. As the planet warms, two particular mosquitoes, *Aedes aegypti* and *Aedes albopictus*, expand their territory. These are the subtypes that transmit the Zika virus as well as the viruses that cause dengue fever, yellow fever, West Nile fever, chikungunya, and eastern equine encephalitis.

Bernard Goldstein, then at the Robert Wood Johnson Medical School, and Donald Reed, of Oregon State University, noted in 1991 that with increasing concentrations of carbon dioxide from the burning of fossil fuels, not only would there be effects of global warming on health, but the concentrations of many serious air pollutants in urban areas that can cause lung disease would also increase.[10] Studies of the linkage between health effects from fossil fuels and climate change were launched in 1997 at a meeting organized by Devra Lee Davis of the World Resources Institute.[11] Particulate matter was chosen as the main topic because it was considered a sentinel air pollutant commonly associated with fossil fuel combustion. This group calculated the global health effects of projected "business as usual" particulate emissions relative to projected reductions in carbon dioxide emissions to 15 percent below 1990 levels by the year 2010 for developed countries and a 10 percent reduction of projected 2010 emissions for developing countries. They predicted that by 2020, seven hundred thousand worldwide deaths could be avoided each year.[12]

Several other studies have more recently examined the comorbidities for health and climate change, but there still appears to be little public understanding of this concept. One of the reports published in 2001 investigated the potential local health benefits of adopting greenhouse gas mitigation policies using readily available technologies to lessen fossil fuel emissions in Mexico City, Santiago, São Paulo, and New York City. They estimated the reduction in health effects from the attenuation of particulate matter and ozone over the subsequent two decades. Following these measures would avoid approximately 64,000 premature deaths, 65,000 chronic bronchitis cases, and 37 million persons' days of work lost or other restricted activity. Economically, the effects of reduction in air pollution would be significant as well. A study by the EPA reported in 2013 that the total economic value of fossil fuel health impacts in the United States from all power plants was an estimated $361.7 billion to $886.5 billion per year.[13]

• • •

Another pollutant from burning fossil fuel is ozone, but the pathway to its production is more complicated. Ozone is a reactive oxygen gas molecule that contains three oxygen atoms (O_3) instead of the two that are found in the life-sustaining form found in air (O_2). Air contains primarily oxygen and nitrogen, and when combustion occurs at high temperatures, some of the oxygen reacts with the nitrogen to form nitrogen oxides. The ultraviolet rays of sunlight then break up the resulting nitrous oxide, which in turn reacts with other oxygen molecules in the air, adding an extra oxygen atom to form ozone. Other products of fuel combustion called volatile organics enhance the formation of ozone in this process.

We now know that ozone in the air that we breathe can destroy the alveoli, the small air sacks in the lung essential for proper respiration.

Similar to the studies of sulfur dioxide using chambers to expose patients, investigators at the University of Washington found that ozone-exposed subjects were more susceptible to the subsequent effects of sulfur dioxide.[14] Observations at the University of Toronto by Nestor Molfino and colleagues showed that on days with higher ambient ozone levels, asthmatics were more susceptible to allergens.[15] Studies done at the Johns Hopkins Bloomberg School of Public Health have found that fracking, which can release volatile organics, is also linked to asthma flareups in Pennsylvania.[16]

In Los Angeles, where I grew up in the 1950s, one only had to take a deep breath to understand that something was hurting our lungs. What we were feeling was ozone, but we knew it as smog. In Los Angeles the main culprit was not the burning of coal; rather it was automobile exhaust, which produced nitrogen dioxide, which in turn facilitated the production of ozone. The burning of natural gas in electric power plants also produces ozone in addition to carbon dioxide. Thus, despite its reputation as the "best" fossil fuel to burn, natural gas contributes both to climate change and to immediate negative health effects.[17] Natural gas combusted in new, efficient natural gas power plants does emit 50 to 60 percent less carbon dioxide compared with the emissions from a typical new coal plant. Also, since natural gas does not produce sulfur dioxide, particulates, or mercury emissions like coal, its use is overall less damaging to health.[18] Oil tends to be between coal and natural gas both in the production of toxins and greenhouse effects.

But new health problems with natural gas have been emerging. The *New York Times* reported that air pollution caused by natural gas production using fracking was a growing threat in rural Wyoming. The state failed in 2009 to meet federal standards for air quality for the first time in its history partly because of the ozone levels resulting from roughly 27,000 wells, the vast majority of which were drilled in the 2000s. The volatile organics, along with the nitrogen oxides from the energy industry's high levels of truck traffic, contributed to ozone levels higher than those recorded in Houston and Los Angeles, despite the state's sparse population. Adding to this, volatile organics are also contained in natural gas, and the geography of the Wind River Mountains in Wyoming traps these gases in much the same fashion as the mountains around Los Angeles. The end result is that the difficulty in breathing in Los Angeles in the 1950s can now be felt in rural Wyoming, where there are only two residents per square mile.[19]

Hydraulic fracturing ("fracking") is a process used to stimulate well production in the oil and gas industry.[20] It is not a new process, but its use has increased dramatically over the last twenty years as technology for the extraction of natural gas improved. Fracking involves pumping large volumes of water and sand into a well at high pressure to fracture tight rock formations, allowing oil and gas to flow into the well. Sand is used in large quantities, up to ten thousand tons of sand per well. Field studies by the National Institute of Occupational Safety and Health (NIOSH) have shown that workers may be exposed to sand with high levels of respirable crystalline silica during hydraulic fracturing operations, leading to the development of silicosis. This is the same lung disease suffered by miners and described in chapter 4.[21]

Another big problem with fracking and other means of producing of natural gas is the release of methane. Methane is the major component of natural gas and a powerful greenhouse gas, being some thirty to sixty times more efficient than carbon dioxide at trapping heat in the atmosphere. All natural gas production leaks methane, often in significant amounts, as do the storage tanks and pipelines that transport it. Some production fields have been found to lose about 4 percent of their gas to the atmosphere even before it enters the pipeline and distribution system. This methane problem upsets some of the calculations for the benefits of natural gas over coal, since coal mining does not release methane in significant amounts.

Getting back to the local effects, we haven't yet mentioned the effects of fracking on groundwater, which can contaminate drinking water from wells. Besides the water and sand injected into the ground, some of the other most widely used chemicals are isopropyl alcohol (rubbing alcohol), 2-butoxyethanol (used to maintain the plastic in the interior of cars), and ethylene glycol (antifreeze). Between 2005 and 2009, the oil and gas service companies used hydraulic fracturing products containing twenty-nine chemicals that are known or possible human carcinogens, regulated under the Safe Drinking Water Act for their risks to human health, or listed as hazardous air pollutants under the Clean Air Act. These twenty-nine chemicals were components of more than 650 different products used in hydraulic fracturing.[22]

• • •

Major air pollutants are controlled by the National Ambient Air Quality Standards (NAAQS). These were first developed in 1971 to put limits on air levels of sulfur dioxide, nitrogen oxides, ozone, particulates, lead, and carbon monoxide. The strictest limits were originally placed on ozone, and since 1971 the regulatory levels have not changed significantly. The allowable levels for sulfur dioxide and nitrogen oxide have been lowered somewhat, but the biggest change in the NAAQS has been in allowable particulate levels, especially in the small particulates.[23]

Particulate matter in the air is a major cause of respiratory diseases, and the small particles are produced by the burning of fossil fuels, especially diesel. The small particles are called PM 2.5s, meaning that they are less than 2.5 microns in diameter, which is small enough to penetrate into the deeper airways of the lungs.[24] These smaller particles also pose a lung cancer risk because they contain PAHs and carcinogenic metals. While the smaller PM 2.5s pose a significant threat, larger particles in the air, called PM 10s, are composed of all kinds of environmental dust and are mostly removed from inhaled air by the upper airways before they can do as much damage to the lungs. This action depends on mucus-producing cells and ciliated cells that trap the particles and cause them to be swept upward so they can be swallowed—not a terribly appetizing thought, but certainly safer than the alternative.

Researchers at Brigham and Women's Hospital and Harvard Medical School investigated hospital admissions in thirty-six cities throughout the United States. They found that short-term changes in particulate and ozone concentrations were related to hospital admissions for emphysema and pneumonia, especially during the warm season.[25] Scientists from Loma Linda University School of Public Health also studied the increase in lung cancer associated with elevated levels of ozone and particulates. Their Adventist Health Study on smog is particularly helpful because good health records are available for people who have uniform and relatively low-risk lifestyle factors, such as no smoking or drinking alcohol and preferably a vegetarian diet. In this study, increased risks of lung cancer incidences among Californians were found to be associated with elevated long-term ambient concentrations of particulates and sulfur dioxide in both genders and with ozone in males. The increase in risk for ozone and particulates was threefold and for sulfur dioxide more than double.[26] A recent study by scientists at Harvard's Department of Environmental Science and School of Public Health found evidence that short-term exposures to PM 2.5s and ozone, even at levels much lower than the current daily standards, are associated with increased mortality, particularly for susceptible populations.[27]

"Clean coal technology" was developed during President Carter's administration in response to the energy crisis and was thought to be a viable way to meet the energy needs of the United States. Besides the usual types of air pollutants, coal also emits "air toxics," which were not covered by the NAAQS but for which some state and local regulatory agencies began setting allowable levels. These included elements such as arsenic, chromium, and nickel and volatiles such as benzene or toluene.[28] By 1990 about ten states had established acceptable levels for mercury. Air models for new power plants had to predict the amounts the plants would release into the air.

Congress amended the Clean Air Act in 1990 to give the EPA the authority to regulate 189 air toxins, including mercury, arsenic, and cadmium. Although the Clean Air Act provided a platform for the regulation of these air toxins by the EPA, it took several years before the EPA proposed levels of allowable mercury emissions. The legality of these standards was then challenged, and the issue rose through the courts. Industry groups lobbying the Supreme Court in this case declared that the government had imposed annual costs of $9.6 billion to achieve about $6 million in benefits. The EPA countered, saying that the costs yielded tens of billions of dollars in benefits. The vastly different calculations of the costs and benefits show how political spin can affect public health decisions.[29] After several court battles, the Supreme Court struck down the regulation in June 2015, by which time most of the environmental controls had been implemented at coal plants across the country. In the courts, however, it was found that the EPA evaluation unjustifiably did not adequately include consideration of cost and required that this be done.[30] At the end of 2018 the EPA announced its proposed rule that drastically revised the estimates of health benefits from decreasing mercury emissions from \$80 billion to about \$5 million, while the costs were about \$8 billion.[31]

• • •

Let's now turn to another means of energy production: nuclear energy, which along with hydroelectric, solar, and wind power, has the distinction of not contributing to global warming. Nuclear power is the greatest source of low-carbon electrical energy generation in the United States and is second in the world only to hydroelectric power.[32] The two are not without their inherent problems. Hydroelectric power does not really have discernible toxicological or health issues. The problems with hydroelectricity generation primarily concern displacement of populations, effects on ecosystems, and accidents during construction.

The main concern with nuclear power is the radiation from accidents, and tragically, a good deal of our knowledge here comes from past accidental contaminations. By studying the most severe radiological accident to date, we can estimate the risk of cancer from the catastrophic failure of a nuclear power plant. The Chernobyl nuclear power plant, in Ukraine, experienced a major accident on April 26, 1986, resulting in the release of several types of radionuclides, including short-lived ^{131}I and long-lived ^{134}Cs and ^{137}Cs. Long-range transport of these and other radionuclides caused serious contamination of Belarus, Ukraine, and the western part of the Russian Federation, along with other parts of Europe. It is considered the largest technological disaster of the twentieth century.[33]

Studies have suggested that by 2006, the Chernobyl accident might have caused about one thousand cases of thyroid cancer and four thousand cases of other cancers in Europe, representing about 0.01 percent of all incident cancers. Models predict that by 2065 about 16,000 cases of thyroid cancer and 25,000 cases of other cancers may be expected from the accident. As we have seen in the previous description of risk assessment for PCB cleanup standards, there are assumptions at play about the extrapolations that are not only related to dose but also applicability to the affected populations. Estimates thus have large degrees of uncertainty. The Chernobyl estimates range from 3,400 to 72,000 for thyroid and 11,000 to 59,000 for all cancers.[34]

These risk estimates do not include the acute exposures of the immediate responders at the Chernobyl site. The horrible details have been documented by Svetlana Alexievich, in *Voices from Chernobyl*, for which she won the Nobel Prize in literature in 2015. The 340,000 military personnel assigned to clean up the reactor site were ill equipped and were exposed to significant amounts of radiation. The worst of these exposures was suffered by the 3,600 workers on the roof of the reactor, charged with ridding it of fuel, graphite, and concrete.

Because the reactor was in danger of exploding like a three-to-five-megaton bomb, which would have spread radiation far enough to render Kiev, Minsk, and much of Europe uninhabitable, men had to dive into a pool of radioactive water to open the bolt on the safety valve so that the water wouldn't enter the compartment of uranium and graphite. In the end, they were successful in preventing the explosion, but as Alexievich writes, "These people don't exist anymore."[35]

Another aspect of toxicity in nuclear power generation is the mining of uranium, which has long been associated with health problems. Radon gas progeny particles are emitted from uranium-rich ore, which attach themselves to natural aerosols in underground mines, causing lung cancer. A study of health costs from excess lung cancer mortality among uranium miners in Grants, New Mexico, the largest uranium-producing district in the United States, found that total health costs ranged from $22.4 to $165.8 million over the 1955–1990 mining period.[36]

For comparison to fossil fuels, it is also worth considering the risks associated with renewable energy sources such as solar and wind power. Wind and solar technology are now in favor thanks to their lack of impact on climate and health. But the development of this technology is heavily dependent on the rare earth metals necessary for components in the wind power turbines. Energy storage technology is also required for renewables, and this uses another rare earth, vanadium, for flow batteries. Hopefully, alternatives such as the iron-containing electrolyte ferrocyanide can someday replace vanadium use.[37]

China mines 95 percent of the global supply of heavy rare earths. Rogue operations controlled by gangs in southern China produce an estimated half of the supply, with legal, state-owned mines mainly accounting for the rest of China's output. It is difficult to get much information about the health risks of this industry because gangs have terrorized villagers who dare to complain about the many tons of sulfuric acid and other chemicals being dumped into streambeds during the processing of ores.[38]

If we examine the manufacturing process of solar cells, solar energy poses similar complications. The vast majority of solar cells start as quartz, which is refined into elemental silicon, and the mining of quartz can cause silicosis. Turning this quartz into metallurgical-grade silicon happens in giant furnaces, and keeping them hot also takes a lot of energy. Next, turning metallurgical-grade silicon into a purer form called polysilicon creates the extremely toxic compound silicon tetrachloride, which one Chinese company dumped on neighboring fields, rendering them useless for growing crops and inflaming the eyes and throats of nearby residents. Silicon is formed into wafers, and manufacturers rely on hydrofluoric acid to clean these. If hydrofluoric acid touches an unprotected person, this highly corrosive liquid can destroy tissue and decalcify bones. In thinking about the cost-benefit analysis here, the occupational risks must be compared to other portions of the energy sector. At the very least, while making solar cells requires a lot of energy, it has been estimated that they do pay back the original investment in terms of energy after just two years of operation.[39]

• • •

The bottom line is that there is no form of energy production that is completely free of health effects on the workers or on the environment, which could affect broader population health. The role of the toxicologist is to analyze all of these energy production methods so that the government and public can make reasonable choices. From what I have found, there is both a health and climate benefit from choosing sources of energy such as wind and solar over fossil fuels. Nuclear energy is an option if one can assure that plants are designed properly and the risk of catastrophic failure such as the one at Chernobyl is very low. As our technology becomes more sophisticated, we should be able to minimize the risks from nuclear accidents and develop wind and solar power utilizing more safely available source materials.

The historical and political context of the debate concerning the best approach for generating power often seems to override the scientific issues. Take, for example, attitudes about nuclear energy before and after Chernobyl and the other accidents at Three Mile Island, in the United States, and Fukushima, in Japan. Before these incidents, nuclear energy was thriving, and had they not occurred, nuclear power would seem far more viable to the general public. There would have been hypothetical concerns about accidents and the storage of spent fuel, but given the lack of air pollution and effects on climate, nuclear energy would seem to be the best solution to our energy needs. Neither Three Mile Island nor Fukushima produced the widespread health effects that occurred with Chernobyl, but they clearly had a major impact on public opinion and political considerations. After Three Mile Island, no nuclear power plants were planned for construction in the United States. And after Fukushima, both Japan and Germany decided to retire their existing plants. But from a toxicology and climate change perspective, this was a questionable decision, since we should be able to learn from these accidents and contain the risks. Even taking Chernobyl into account, the cancers and deaths from nuclear power are minuscule compared to the health effects from fossil fuel production and use.

To summarize, fossil fuel use is the primary cause of premature deaths in the world and the primary cause of climate change. According to the World Health Organization (WHO), air pollution from burning fossil fuel and climate change top the list of global health threats. They estimate that climate change is responsible for 250,000 deaths per year in the world from malnutrition, malaria, diarrhea, and heat stress. But this pales in comparison to the 7 million people killed prematurely by air pollution per year from cancer, stroke, and heart and lung disease. Worldwide ambient air pollution accounts for 29 percent, 24 percent, 25 percent, and 43 percent of all deaths from lung cancer, stroke, ischemic heart disease, and emphysema, respectively.[40] The head of the WHO said in 2018 that air pollution is the "new tobacco."[41] Our energy policy should take into account the health impacts of our energy consumption, including the health impacts of climate change.

24

Animal Models
for Human Disease

A s discussed throughout this book, in studying disease, there
has been a historic tension between the study of humans and
the study of animals. The studies of humans by Paracelsus
and Ramazinni have been continued by modern-day epidemiologists, and these studies are directly relevant to humans. On
the other hand, animal studies have been important in helping
us identify and understand the toxic principles in mixtures as
diverse as venoms and soot. However, given the vast amount of
information that research using experimental animals has
amassed, there remains the question of how much the results
enable us to understand human disease. To use the rodent model,
we must have faith that humans, mice, and rats all respond in
the same way to chemicals. But are rodents similar enough to
humans to be good predictors of chemically related disease vul-
nerabilities? Cancer and other diseases are extremely complex
and genetically driven, which may make them more species specific than the regulatory community might admit.

Genetic analysis reported in 1998 disclosed that primates and
rodents began evolving on separate paths about 100 million years
ago. After that time, further evolution of primates led to humans,

and rats and mice later evolved from ancestral rodents.[1] However, basic differences among the genomes of rodents and humans are even further amplified by downstream differences in the "wiring" between the various gene products that control cell cycle regulation.[2] These in turn create significant differences in cell division and growth regulatory pathways, indicating that we might not expect test results in rats and mice to prove or disprove whether a chemical will cause cancer or other diseases in humans. An emerging body of evidence indicates that there are fundamental differences in how the process of tumorigenesis occurs in mice and humans.[3]

A big hint about this difference is in the types of spontaneous diseases and tumors we see in rodents and humans. One crucial example of this is chronic progressive nephropathy (CPN). This kidney disease is very common in rats of strains used in bioassays, but there is no disease entity in people that has the singular features of CPN. In the United States, diabetes is the most common cause of human renal failure, while in Australasia the primary cause is glomerulonephritis. Postinfectious glomerulonephritis is still the most prevalent single cause of end-stage renal disease in developing countries. Together, diabetes, hypertension, and glomerulonephritis make up three-quarters of end-stage renal disease in humans, but in rats it is CPN.[4]

What about spontaneous cancer rates in humans compared with rats, that is, cancers generated by the aging of cell tissue rather than by a specific chemical or environmental cause? For the male Fischer 344 rats, which are the rats used by the National Toxicology Program until 2006, 90 percent get testicular tumors, 50 percent mononuclear leukemias, 30 percent adrenal gland tumors, and 30 percent pituitary gland tumors by the end of their lifespans. All of these are uncommon or rare neoplasms in humans. In B6C3F1 mice, which are the mice still used by the National Toxicology Program, the most frequently occurring neoplasms are liver tumors, which in the United States are much less common in humans, although that is not true for some parts of the world.[5]

• • •

An investigation in 2004 presented evidence that human disease genes differed significantly from rat and mouse genes depending upon the system and type of disease. Rodent and human genes associated with neurological function exhibited the greatest evolutionary similarity, and this suggests that rodent models of human neurological disease are likely to represent human disease processes most faithfully. In contrast, the systems with the least genetic similarity were those of the immunological system, followed by the hematological, pulmonary, and hepatic-pancreatic systems.[6] One of the reasons for the immunological system being the least similar between humans and rodents is that the selective pressures that normally result in evolution, and ultimately in speciation, occur as a result of adaptation of organisms to their environment. One of the major components of an organism's environment is the pathogens to which it is exposed, putting pressure on the immune system to evolve and keep pace with infections. This concept forms the basis of the "host pathogen arms race" and offers a reason why animals may not be good models for studies of the immune system.[7]

The metabolism of chemicals may also be different in different species. A study of species differences across mouse, rat, dog, monkey, and human drug metabolisms concluded that the species-specific isoforms of the liver's drug metabolizing enzymes show appreciable interspecies differences in terms of their activities. Therefore, caution should be applied when extrapolating metabolism data from animal models to humans. In contrast, CYP2E1, which metabolizes many industrial chemicals, shows no large differences between species, and extrapolation between species appears to hold quite well. This is good news for rodent studies because metabolic activation by CYP2E1 leads to cancer for many industrial chemicals and ethanol. For example, the similarity between rats and humans in angiosarcoma of the liver caused by vinyl chloride even extends to the dose required to show significant numbers of tumors.[8]

We also need to be careful and understand the similarities and differences in the spectrum of cancers caused by germ-line mutations between humans and rodents. Evaluation of mouse models of inherited cancer susceptibility syndromes were carried out by William Hahn and Robert Weinberg, of the Whitehead Institute for Biomedical Research and affiliated with the Massachusetts Institute of Technology. Building on the work of Michael Bishop and Harold Varmus, who were awarded the Nobel Prize for their demonstration that oncogenes were derived from cellular proto-oncogenes, Weinberg was one of the discoverers of the effects of oncogenes in human cells. Hahn and Weinberg's study shows how this molecular circuitry of genes makes a big difference between species. Their examination of *p53* tumor suppressor gene–inherited mutations demonstrated that brain cancers, breast cancers, and leukemias derived from this gene mutation are found in humans but not in mice. In contrast, mice whose *p53* gene has been genetically modified suffer from lymphomas and soft tissue sarcomas. The *p53* gene is important in preventing the development of cancer because it halts cell replication when there is too much genetic damage, thereby preventing mutations. Other differences between mice and men involve the gene networks surrounding the *p53* gene. Although *p53* is mutated in 50 percent of human cancers, humans do have some pathways redundant to the *p53*, which offer further protection. Mice lack these redundant pathways, so for them even one mutational event might be sufficient to cause cells to replicate indefinitely, incorporating genetic damage that will go unrepaired. In additional to the *p53* gene, other inherited gene mutations also have specific patterns of cancer risk. For the inherited syndromes involving the retinoblastoma (*Rb*) gene in mice, there are brain and pituitary tumors, whereas in humans there are retinal and bone tumors. For mutations of the *BRAC 1* and *BRAC 2* breast cancer susceptibility genes in humans, there are no counterparts in mice.[9]

Cancer researchers can create immortal cells in the laboratory by freeing normal cells from their usual restraints so that they continue to divide in culture without dying or halting cell division. Hahn and Weinberg also made the observation that rodent cells are more easily transformed into immortalized cells than human cells. Normal cells will divide a few times and then stop, but cancer cells divide continuously and are thus said to be immortal. Human cells must undergo at least four to six mutations to become immortal, but a mouse cell may need only two. In fact, it has been very difficult to create immortalized human cells in culture and relatively easy to develop immortalized mouse cells just by repeatedly passing them from one culture medium to another.

One of the reasons for this species difference has to do with telomeres, which are repetitive DNA sequences at the ends of chromosomes that protect them and regulate the lifespan of cell lineages. Since DNA replication cannot continue to the very end of the chromosome, telomeres become shorter with each cell division. In order for cells in humans to divide continually, as happens with cancer cells, they must surreptitiously acquire telomerase activity, which is an enzyme that adds to the end of the DNA sequence and thus maintains telomere lengths. However, inbred mice used in animal models do not need to acquire telomerase activity because they have telomere lengths three to ten times greater than those of human cells. Consequently, they are much more prone to cancer development.[10]

Besides cancer, other types of toxic effects are also species dependent. A multinational pharmaceutical industry survey has attempted to produce a concordance of the toxicity of pharmaceuticals observed in humans with that observed in experimental animals. The main aim of this project was to examine the strengths and weaknesses of rats, mice, dogs, and monkeys to predict human toxicity. Not surprisingly, monkeys were the best, followed by dogs, then rats. Mice were the worst in terms of predictability of human toxicity.[11] And it's not just toxicity issues for pharmaceuticals that are a concern, it's also efficacy. Although mice are widely used in pharmaceutical research, questions remain about their reliability as a model for treating human diseases. Many drugs work well in preclinical trials in mice but turn out to be ineffective when used in clinical trials on humans.[12]

• • •

At the American Health Foundation, epidemiology studies by Ernst Wynder had shown that diets relatively high in fat were associated with increased risks of breast cancer. Using chemical-induced mammary cancers in rats, other investigators at the foundation studied the effects of different fats on cancer development. I was investigating chemical carcinogenesis at the foundation during the time that these breast cancer studies were being done. So when I got to know Dr. Larry Norton, chief of the Division of Solid Tumor Oncology in the Department of Medicine, specializing in breast cancer, at Sloan-Kettering, I raised the issue of dietary fats. I described to him the research at AHF regarding dietary fat in experimental animals and which types of fat, such as certain unsaturated fats, should be considered "bad" for breast cancer from the results of their studies. I'll never forget his abrupt response: "We don't have long naked tails."

At first I was taken aback, and I interpreted his attitude as unscientific. Here was one of the most respected researchers in breast cancer treatment dismissing the experimental research studies at the AHF as irrelevant when it came to humans. Later, when I had had time to think about it, I realized that what Norton was expressing was a clinical researcher's skepticism of animal studies. Larry Norton is a "trialist," a term for a clinical researcher who is experimenting on human cancer in clinical trials. His focus is on killing human cancer cells because that is how one cures cancer once it develops.

His remark made me realize that the use of chemical-induced animal models to study human cancer has largely fallen by the wayside compared to other methods, partly because the ability of animal models to predict which drugs cure cancer in people is often disappointing. My colleagues and I lacked Norton's gritty pragmatism. We were enamored with our laboratory studies. We clung to the animal model because it provided us with clear-cut experiments to test a hypothesis that would give us an answer—but only an answer in rodents. Later the studies of types of fats and breast cancer risks were not verified in humans, thereby confirming Norton's suspicions about animal models. If anything, saturated fats may contribute to breast cancer risk.

One of the most serious obstacles facing investigators involved in the development and assessment of new anticancer drugs is the failure of rodent tumor models to predict reliably whether a given drug will have antitumor activity and acceptable toxicity in humans. Do researchers use chemical-induced cancers in experimental animals as models to test cures for cancers in humans? The answer depends on the cancer type and the chemical. For example, diethyl nitrosamine produces liver tumors in rodents that have some of the same genetic inactivations found in human liver cancer. The lung cancer model in mice induced by urethane has found use in studying human lung cancer.[13] However, chemical models have not proven themselves for many types of cancer, including small cell lung cancer in humans.[14]

The methods now used to study human cancers are different from the traditional rodent bioassays that toxicologists have used to sift out "cancer-causing" chemicals. For example, oncologists have not used PCB-induced liver cancer to study chemotherapeutic agents for the treatment of human liver cancer. More commonly, oncologists use animals that have been transplanted with tumors that originated in humans. Or they use experimental animals that have been genetically engineered to develop cancers more closely resembling human cancer. The genetic alteration in these mice, so-called transgenic mouse models, often involve the insertion of genetic material in a cell that will become an embryo that inactivates a particular gene such as *p53*. It is too early to conclude, one way or the other, whether these will be superior to transplantable human tumor models.[15]

In 1971 Judah Folkman, at the Children's Hospital Medical Center and Harvard Medical School, reported that human and animal solid tumors elaborate a factor, called tumor angiogenesis factor, that causes capillary cells to grow in tumors. Folkman suggested that blockade of this factor (inhibition of angiogenesis) might arrest the growth of solid tumors.[16] In 1980 Folkman showed that angiogenesis inhibitors called angiostatin and endostatin occur naturally and in small quantities in the human body. They limited tumor growth in mice by stopping the formation of blood vessels. Furthermore, the inhibitor appeared to have no toxic effect directly on tumor cells and no short-term toxicity for the host.[17] Folkman found that all tumors responded to the drugs in the same way. Even leukemia responded because it needs to form new blood vessels in the bone marrow to grow. Folkman was cautiously optimistic about this treatment approach, saying, "If you have cancer and you are a mouse, we can take good care of you."[18]

However, after thirteen years of trials in humans, the only beneficial effect of endostatin reported was a two-month increased survival in human lung cancer patients. For other cancers such as breast cancer, no effects were found based on the mouse studies.[19] So initially there was much disappointment in the medical community that Folkman's silver bullet for cancer, so promising in mice, was virtually ineffective in humans. Eventually there were signs that some other cancers responded to antiangiogenesis therapy, but it was certainly not the panacea Folkman had hoped for.[20] The problem with the success rate in humans was that advanced stages of human cancers such as breast cancer can express up to six proangiogenic proteins. In neuroblastoma a high expression of seven angiogenic proteins is found, and human prostate cancer can express at least four angiogenic proteins. Dealing with the multiple proteins is much more complex than the angiogenic responses in mice.[21]

In Australia in 2003, after more than three decades of research on angiogenesis, the translation of antiangiogenic therapy from the laboratory to the clinic was finally underway. Thalidomide, the same drug that had caused major birth defects in children in the 1950s, was the first antiangiogenic drug in Australia to get approval. Thalidomide treats multiple myeloma, which is a type of leukemia involving lymphocytes that produce antibodies. Next in 2004 and 2005, Avastin was approved in the United States and twenty-six European countries to treat colorectal cancer. Others followed for treatment of macular degeneration and lung cancer.[22]

• • •

On top of the underlying differences between animals and humans, animal studies for pharmaceutical development have come under scrutiny because of their lack of reproducibility. Scientists from the biotech company Amgen reported in 2012 that they were able to reproduce only six of fifty-three high-profile cancer research papers. The pharmaceutical firm Bayer reported a 79 percent failure rate for reproducibility of preclinical cancer studies. Amgen argued that irreproducible animal studies contributed to high drug development costs and failed clinical trials.[23] Part of this failure could be attributable to differences in strains or even the batches of mice used in the studies. Causes of different experimental results can result from environmental variables in the animal facility, including temperature, type of bedding, shelf level, and the time of day the drugs are administered in the experiment.

Attempts have been made to solve this problem. In 2010 the UK National Centre for the Replacement, Refinement, and Reduction of Animals in Research developed guidelines for information to be included in animal research publications. These ARRIVE guidelines—short for "Animal Research: Reporting of In Vivo Experiments"—were endorsed by over one thousand scientific journals and two dozen funding agencies. However, most researchers have either ignored or remain unaware of the guidelines. The items that need the most improvement are randomization of animals between experimental groups, blinding of researchers to experimental group identification, identification of why any animals might have been dropped from the study, and accurate calculations of the sample size of groups necessary for the experiments.[24] Without compliance with these guidelines, bias in the conduct and reporting of the results of an animal studies can arise. The adherence to similar rules is common practice in human experimental and epidemiological studies, which is a fundamental reason why they can be more reliable.

Another problem with animal study reproducibility is related to the differences in the bacterial populations in the intestines of mice, otherwise known as the microbiome. This has been found to hinder the reproducibility of studies even when the mice are of the same strain and obtained from the same source. Laura McCabe, at Michigan State University, found in her first experiments that the administration of new drugs to mice caused bone loss. But then she repeated her study with mice from the same source and under the same conditions and found bone density gain. In yet a third experiment, she found no change. The explanation was found to be three distinct mouse microbiomes in the three experiments.[25]

Robert Perlman, at the University of Chicago Medical School, summed up the situation regarding mouse models of human disease as follows:

> Unfortunately, despite the many attempts to translate the results of mouse research to humans, we still cannot specify in advance which research in mice is likely to benefit or shed light on human biology and health. For the most part, we have only anecdotal information about studies in mice that translated to humans and those that didn't. We need more systematic collection, reporting, and analysis of mouse research (and research on other "model organisms") to figure out what works and what doesn't. Until we have that information, we need to be more critical in pursuing mouse research and in making claims about the applicability of this research to humans.

Like Larry Norton, Perlman is comparing the results of research in rodents to human disease as a clinician and researcher. As we will see in the next chapter, the same could be said for the use of rodent bioassays to predict human cancer.[26]

25

Are Animal Cancer Bioassays Reliable?

When we hear that a chemical has been labeled a "carcinogen" or as "cancer causing," the connotation is often that it is imminently dangerous and proven to cause fatal disease in humans. But the term "carcinogen" usually does not describe risk that is so cut and dried. Often, something has been labeled a carcinogen because it causes cancer in rats or mice—rats or mice that have been fed huge amounts of chemicals from the age of eight weeks until their dying day. We saw in the previous chapter that rats and mice have some distinct differences in evolution, metabolism, genetics, and molecular circuitry compared with humans. As a result, rats and mice have different spontaneous cancers and would be expected to react differently to chemicals. Not all chemicals that are labeled a "carcinogen" are proven to cause cancer in humans.

Even though the number of drugs that have been found to be proven human carcinogens by the IARC is only about twenty, our government requires pharmaceutical companies to conduct extensive rodent bioassays for carcinogenicity on every product before it will grant them approval to market their products, and the government in turn must review these tests. Synthetic chemicals are likewise tested extensively by both the chemical industry and the government, even though only about thirty have been classified as human carcinogens. On the other hand, there are about three hundred synthetic chemicals in lesser categories, mostly because of findings in animal bioassays. Despite all of this animal testing and regulatory activity, have cancer rates improved dramatically? The answer is no—with the exception of those that can be explained by other factors, such as lower smoking rates and lower exposure to some chemicals in the workplace. But my argument is not against animal testing in general. The judicious use of animals has been fundamental to the development of toxicology.

Let's focus first on the development of bioassays in rats and mice to predict the ability of chemicals to cause cancer in humans. Although the rodent bioassay was designed by the Weisburgers at the National Cancer Institute forty years ago, the basic design has not changed. A chemical is administered over their lifetime to an experimental group of rodents—traditionally rats or mice—to see if they develop more tumors or develop tumors at a faster rate than a control group does. Since it is prohibitive to use large enough groups of animals to show a statistical difference at a human dose, we usually compensate by using very high doses of the chemical, called the "maximum tolerated dose," in groups of fifty animals per sex and per dose. The bioassay is usually run by giving the chemical in food or drinking water, and exposure frequency depends on the eating and drinking habits of the rodents. Some volatile chemicals have been given by inhalation exposures, often for eight hours per day, five days per week.

The rodent cancer bioassay was first adopted as a way of studying human cancer because almost all human carcinogens had been found to cause cancer in rodents. However, since most human carcinogens are genotoxic, these results pertained to a cancer mechanism common to both humans and rodents, that is, damage to DNA. David Rall was the first director of the National Institute of Environmental Health Sciences, from 1971 until 1990, and was also director of the National Toxicology Program from its beginning in 1978. He reported an unpublished study in 1979 by Bernard Altschuler, of New York University, of the 337 materials reported in the first thirteen IARC monographs. He found that, of the eighty-one that had positive evidence for cancer in animals, twenty-two of these showed "definite" and fifty-nine "less definite" evidence of carcinogenicity in humans, leading to his conclusion that the sensitivity of the screen was 98 percent and the specificity 100 percent.[1]

In contrast, as of this writing, there are about sixty single chemicals (including elements) that have been officially identified as human carcinogens by the IARC, and these are split between industrial chemicals, pharmaceuticals, and naturally occurring products.[2] Most of these were also found to cause cancer in rodents, and most are genotoxic, with others acting via immunosuppression or estrogenic effects.[3] In most cases identification of the chemical as a carcinogen was made in humans first and only later confirmed in rodents, sometimes with difficulty. For example, up until 2004, the IARC still considered the evidence for arsenic in experimental animals to be insufficient, given the lack of clear bioassay tumor findings.[4] It was not until 2012 that the IARC decided that there was enough evidence in experimental animals; however, the inhalation exposure of gallium arsenate produced increased lung tumors in female but not in male rats or in mice.[5]

It may be that most human carcinogens will cause cancer in animals because they are usually genotoxic, but it is a big step over a logical divide to assume that animal carcinogens, which are often not genotoxic, necessarily pose a risk to humans. The validity of the animal model appears to be assumed, rather than critically analyzed in a manner similar to the Hill method used for epidemiology studies. Yet the scientific and regulatory community has persisted in having faith in this model, even in the face of evidence that it hasn't been validated and may be faulty. Some of this was explored in chapters 17 and 18.

A look at differences in cancer bioassay results between rats and mice is revealing, even though those two animals are more closely related to each other evolutionarily than rodents are to humans. When two bioassays of a given chemical are performed in the same way in rats and in mice, the results show the same type of cancer in mice and rats under 50 percent of the time. In 1993, the scientists Joseph K. Haseman and Ann-Marie Lockhart, at the National Institute of Environmental Health Sciences, looked at the results of 379 bioassays performed by either the National Cancer Institute or National Toxicology Program and found that the chances that a chemical being carcinogenic at a particular site in rats will be carcinogenic at the same site in mice (and vice versa) was only 36 percent.[6]

A comparison of the human and rodent genomes confirms the obvious—the differences between us and them are significantly greater than those between rats and mice. The genomic sequence data confirms that the rodent lineage split 12 to 24 million years ago into the separate lines that gave rise to the rat and to the mouse; by contrast, humans and rodents split about 80 million years ago.[7] Therefore we should expect that there would be even less than 36 percent predictability for human cancers from rodent studies. Assuming that rodent cancers predict human cancers until proven otherwise acts on faith rather than on science.

Tumors that are uncommon (or nonexistent) in humans but are often increased by chemicals in rodent sites include those of the testes and thymus plus three sites that humans don't have: the forestomach, the Zymbal gland, and the Harderian gland. These appear against already high spontaneous incidences of mononuclear leukemias, testicular cancer, mammary cancer, and liver cancers in rodents, depending highly on whether we are talking about rats or mice and which strain (subspecies) is being used. Likewise, rodents do not have high normal rates of some common human cancers, such as those of the prostate and colon.[8]

Bruce Ames, the leading researcher of mutagenicity testing, and his coworker Lois Gold found that approximately *half* of all 350 synthetic industrial chemicals tested with a standard high-dose rodent bioassay were found to be rodent carcinogens. They also reported that even though only seventy-seven naturally occurring chemicals from plants have been tested, half of these were also found to be rodent carcinogens. Since many of these are in food, does this mean that the typical human diet is carcinogenic? No, foods such as vegetables and fruits have been shown to promote health.[9] Ames and Gold attributed the high rate of positive bioassay results to the use of toxic high doses. And another problem with the reliability of the bioassay is the lack of replication between studies. Researchers in Austria's Institute for Cancer Research studied the differences between carcinogenicity bioassays reported by the U.S. National Cancer Institute or National Toxicology Program and those published in the peer-reviewed literature. They found that there was only a 57 percent concordance between these two sources of bioassay results for classifying chemicals as carcinogens.[10]

• • •

Risk assessors and pharmaceutical companies are betting people's health on the results of animal bioassays. This gamble does not account for the definite negative consequences of "false positives" from these tests. For instance, if a promising drug for the treatment of heart disease is found in a rodent test to cause tumors, it may be kept from the market even though patients would benefit from it, with lifesaving consequences. This almost happened with reserpine, the derivative of the rauwolfia plant, a widely used early hypertension drug. It increased the incidence of mammary and seminal vesicle tumors in female and male mice, respectively, and adrenal gland tumors in rats. Fortunately, in 1980 the IARC gave the drug a pass and concluded that the evidence found by the National Toxicology Program tests in animals was only "limited."[11]

In another break with past classification methods, the IARC in 1994 decided to classify the cholesterol-lowering drug clofibrate, reasoning that the three studies showing liver tumors in rats did not prove that it was a human carcinogen.[12] These tumors were found to be associated with a proliferation in liver cells of peroxisomes, which are intracellular organelles that perform oxidation of certain lipids. This proliferation occurs only in the rodent liver. When human liver cells are bathed in these drugs, their peroxisomes don't proliferate.[13] There is a long list of drugs and chemicals that produce liver cancer in rodents involving peroxisome proliferation similar to that found for clofibrate. These include other hypolipidemic drugs, which are important for lowering cholesterol, as well as several widely used industrial chemicals. Also, insofar as there are studies available, liver cancer is not produced by these agents in humans.[14]

Researchers had found that the naturally occurring citrus oil constituent limonene causes kidney tumors only in male rats, not in female rats and not in male or female mice. Limonene has been widely used for more than fifty years as a flavor and fragrance additive in perfume, soap, food, and beverages. It is the main volatile and fragrant constituent of the peels of oranges, lemons, and grapefruit—hence its name. Limonene has also been used medicinally for the treatment of gallstones and has been investigated for its ability to shrink tumors of the breast, skin, liver, lung, and forestomach in rodents.[15] We now know the mechanism by which limonene causes the kidney tumors only in the male rat. These tumors are heralded by a particular type of toxicity in the rat kidney called hyaline droplet nephropathy, which is characterized by appearance of reddish droplets in the cytoplasm of the proximal renal tubular cells. In male rats, these cells take up some of the protein alpha 2u-globulin from the urine and catabolize it to amino acids. Limonene causes the massive accumulation of this protein in the cells, which is responsible for the hyaline droplets. Normally, a chemical sex attractant, or pheromone, is bound to the protein alpha 2u-globulin when it is excreted into the urine of male rats. Afterward, the volatile pheromone is released from the protein and signals the female. But if the wrong chemical—limonene or certain other chemicals—is bound to the alpha 2u-globulin protein instead of the normal pheromone, the bound globulin will build up in the renal tubular cells, thereby creating the kidney toxicity that leads to tumors. Limonene cannot cause kidney cancer in the absence of the extensive accumulation of the alpha 2u-globulin protein in the liver, which is specific to male rats. Humans do not have this protein.

One of the most controversial animal carcinogens is saccharin. Saccharin has been used as an artificial sweetener for over one hundred years, but its popularity increased dramatically after the banning of cyclamates in 1970. It was one of the first chemicals to be tested using the rodent bioassay developed by the FDA in the 1940s and published in 1951.[16] It was tested at doses of up to 5 percent of the diet for two years, which is close to the limit of the rat's lifetime, yet no increased levels of tumors were detected. But then the FDA recommended that animals begin exposure while still in their mother's womb (in utero), and this study found a low incidence of bladder tumors in the offspring.[17]

Under the Delaney Amendment to the Food and Drug Act, which banned any chemical causing tumors in animals as a food additive, FDA proposed a controversial ban on the use of saccharin in 1977. It was strenuously opposed by diabetics and the physicians who treat them, as saccharin was indispensable for restricting sugar consumption. At the time, saccharin was the only artificial sweetener that was heat stable and could be used in baked goods. In addition, a dozen epidemiological studies had been done, all of which were unambiguously negative for bladder cancer. So Congress took the unusual step of exempting saccharin from the Delaney Amendment for five years pending new regulations. This exemption was then renewed every five years, until the Delaney Amendment was revoked in 1997. Thus saccharin was never removed from the market.

However, saccharin was still considered to be a carcinogen in animals and of possible relevance to human cancer. Samuel Cohen, at the University of Nebraska Medical School, demonstrated that saccharin produced microscopic stones, or crystals, in the rat bladder, and these crystals were an irritant that ultimately caused tumors.[18] The very high dose of sodium saccharin given to the rats for their lifetime alters their urine chemistry to produce the crystals, which was not produced by the acid form of saccharin; therefore, it was the large amounts of sodium and not just the saccharin that was causing the tumors. Sodium chloride, ascorbate (Vitamin C), sodium bicarbonate, and other chemicals were also found to cause these tumors. The mechanism was the irritation caused by the stones. Scientists at Nagoya City University Medical School found that mice, hamsters, and guinea pigs exposed to saccharin also did not show increased cell division in the bladder; therefore, this effect appears to be unique to the rat. As a result, the bladder tumors caused by saccharin were eventually deemed to be not relevant to human cancer by the IARC and the National Institute of Environmental Health Sciences.[19]

The only chemical that has been found to produce significant amounts of liver cancer in both humans and experimental animals is aflatoxin, a poison from a fungus that can grow on grains and nuts that have not been stored properly. Aflatoxin-induced liver cancer is found in certain Asian and African countries, especially China, where aflatoxin levels are much higher in many types of grains. Liver tumors are only the fourteenth most common tumor in humans in the United States.[20] The major risk factors for human liver cancer in the United States are hepatitis B and C and high, chronic levels of alcohol.[21] Yet liver tumors are the most common cancer caused by chemicals in rodent bioassays.[22] This discrepancy can be largely explained by the fact that chemicals are usually administered to rodents in bioassays by the oral route, and the very high doses used cause chronic toxicity in the liver. At the lower doses of typical human exposures, toxic effects don't occur, and therefore the type of cancer mechanism originally described by Virchow involving toxicity doesn't occur.

Phenobarbital is a drug that has been widely used for epilepsy for decades; it causes benign liver tumors in rats and malignant liver tumors in mice. The connection between phenobarbital and liver tumors seemed to receive epidemiological support in a study led by Jorgen Olsen, of the Danish Cancer Society. Epileptics treated with phenobarbital were found to have an increased incidence of liver cancers. However, at the American Health Foundation, we determined that the increases were caused not by phenobarbital but by the radioactive brain-imaging material Thorotrast used to locate the lesion in epilepsy. It contained radioactive thorium, which was known to cause liver cancer. When the small number of patients who had been exposed to thorium were eliminated from the study, there was no increase in liver tumors over background rates.[23]

Another chemical causing liver tumors in rodents is chloroform, which was once used as an anesthetic for surgery, an organic solvent, and a dry-cleaning spot remover. It is formed at very low levels when drinking water is chlorinated to protect against pathogens. High doses of chloroform administered by gavage (force feeding) in corn oil produce liver tumors via a toxicity mechanism. However, the experimental evidence is convincing that low doses of chloroform do not cause any harm because the toxic process it induces is inhibited by glutathione, an antioxidant normally found in cells. Analysis of the high-dose studies revealed that the tumors were an artifact produced by giving the chloroform to the rats in a single bolus of corn oil each day rather than continuously in drinking water, an exposure regimen that produced no tumors.[24] The policy of the EPA is that "chloroform is not likely to be carcinogenic to humans by any route of exposure under exposure conditions that do not cause cytotoxicity and cell regeneration."[25] Therefore, small amounts of chloroform resulting from chlorination to protect from infectious agents can be allowed in drinking water.

• • •

These studies of clofibrate, limonene, saccharin, phenobarbital, and chloroform show that we need to begin tackling the problem of the reliability of animal bioassays. Although researchers have been studying the mechanisms of chemical-induced cancer development in rodents for several decades, there has not been a systematic approach by those with the most experience, that is, the NTP and FDA, to verify whether cancer is produced in animals in the same way as in humans. Where it exists, which is somewhat piecemeal, cancer mechanism data from rodent studies can be used to show how cancer might be caused in humans even when data in humans is lacking or inadequate. This will be a tricky endeavor. It is basically the mirror image of the problem explored earlier, where working groups of the IARC used mechanistic data to demonstrate that some animal cancers do not apply to humans. More recently, other organizations, such as the International Life Sciences Institute, have also prepared guidelines to determine the relevance for humans of animal bioassay data.[26]

There have been few instances where bioassay methodology has been changed because the results have not been validated. To its credit, the NTP stopped using the F344 rat for cancer bioassays in 2006 because of its very high and variable spontaneous rates of the unusual mononuclear leukemias and Leydig cell testicular tumors as well as the development of another tumor involving the tunica vaginalis of the scrotum. The NTP decided that the variability of the spontaneous tumor rates was difficult to distinguish from increased incidences that could be caused by chemicals. It also considered that the mechanism of increases in these tumors is different from that which would occur in humans.[27] Despite the NTP's informed decision regarding one specific rat species, we are still left in most cases with the results of bioassays that were performed on these rats for over five decades, and there is little guidance on the use of these results in cancer classification. This exemplifies only one of the reasons that there are ten times more established rodent carcinogens than human ones, according to the IARC's classifications.

Proving the positive, that is, that certain animal data is relevant to human cancer, may be more difficult than proving the negative, but since the stakes are so high, it should be tried. Right now, there are two large bodies of information out there—one concerns the development of human cancer, and the other informs us how chemicals cause cancer in rodents. A large amount of money and effort has been invested in each of these data sets, but the amount of research aimed at determining the relevance of one body of information to the other has been minuscule in comparison.

26 | Hormone Mimics and Disrupters

D uring the obstetrics service, a medical student may deliver as many as thirty babies and assist with many more. For many students, this is the best part of medical school; the thrill of helping new life enter the world is almost a religious experience. Of course, medical students are usually assigned to women giving birth who have previously had at least two uncomplicated births and when it has already been determined that the fetus is in a good position in the birth canal. Those women usually know a lot more about birthing than the students.

Medical students also learn about the complex system of positive and negative controls whereby the reproductive system is carefully orchestrated to perform properly. At the center of this control is a part of the brain called the hypothalamus and the pituitary gland, which is outside the brain but located nearby. One of the most important functions of the hypothalamus is to link the nervous system to the endocrine system via the pituitary gland. It performs this feat by producing releasing factors that are directly secreted into the pituitary through small blood vessels. These factors cause the release from the pituitary of follicle stimulating hormone (FSH) and luteinizing hormone (LH), which control estrogen and progesterone secretion by the ovaries. Other hormones produced by the pituitary stimulate the uterus during childbirth and the breasts to produce milk, but for release of these there is a direct neural connection from the hypothalamus to the pituitary.

The ovaries produce estrogens and progesterone from cholesterol through a series of biochemical reactions involving various cytochrome P450 enzymes, the same types of enzymes that metabolize chemicals in the liver. Androgens such as testosterone are produced first and then converted to estrogens by the P450 enzyme aromatase. During the menstrual cycle, the increasing amount of estrogen produced by the ovary stimulated by the FSH eventually causes a positive feedback on both FSH and LH. This "LH surge" releases of the egg from the follicle, whereby the follicle is transformed into the corpus luteum, which produces progesterone. If the egg is fertilized by a sperm, the luteal cells continue to produce progesterone under control of pituitary LH to support the pregnancy. Both progesterone and estrogen levels rise dramatically during pregnancy; estrogen stimulates the milk duct system to grow and differentiate, while progesterone stimulates the glandular cells that produce milk to develop in the breast.

Chemicals called "hormone mimics" or "hormone disrupters" can alter the normal physiology of the female reproductive system either by acting like estrogens or by antagonizing the effects of estrogens. Drugs such as birth control pills and hormone replacement therapy for menopause that contain estrogen and progesterone are hormone mimics. Both types of drugs can be human carcinogens because they stimulate the proliferation of stem cells, which can lead to the development of breast and endometrial cancer. On the other hand, antiestrogens such as tamoxifen can treat or prevent the development of breast cancer.

Our understanding of the biology of female reproduction was advanced at the beginning of the twentieth century. Josef Halban and Emil Knauer, who worked in different gynecological clinics in Vienna, independently transplanted rabbit and pig ovaries in 1900 and concluded that these organs stimulated uterine growth by secretion of a biologically active substance into the blood. Francis Marshall and William Jolly in 1905 hypothesized that ovarian cells secreted a compound that induced estrus in rodents.[1] In St. Louis, Edward A. Doisy prepared the ovarian hormone "Folliculin" from the urine of pregnant women in 1929, from which he later purified estrone and estradiol.[2]

A U.S. woman's lifetime risk of developing breast cancer is one in seven, which makes it a major public health issue, not to mention a source of great anxiety for all women. Interest in hormone mimics was established through studies that showed that estrogen administration produced breast cancer in mice. In Paris, Antoine-Marcellin-Bernard Lacassagne induced mammary cancers in male mice via the injection of a preparation of folliculin from female urine. In 1938 Lacassagne produced mammary carcinomas in male mice using a new powerful synthetic estrogen, diethylstilbestrol (DES), which had been synthesized in London by Edward Charles Dodds.[3]

From 1938 until 1971, U.S. physicians prescribed DES to pregnant women to prevent miscarriages and avoid other pregnancy problems. Its use was halted when DES was linked to a rare type of vaginal cancer in the female offspring of treated women. In 1970 Arthur Herbst and Robert Scully, at Massachusetts General Hospital, first described seven cases of a rare clear cell carcinoma of the vagina in young women between the ages of fifteen and twenty-two. The seven cases exceeded the total number of these cancers in this age group previously reported in the world literature. These clinical investigators discovered that in seven of the eight patients their mothers had received DES during pregnancy for threatened abortion or for prior pregnancy loss. Eventually, over two hundred such cases were reported.[4]

• • •

Although early studies had found that estrogens were capable of inducing mammary tumors in mice, as of 1979 the specific role of estrogens in the development of human breast cancer was still unclear. The administration of estrogens (including diethylstilbestrol) to adult women was first found to be causally associated with an increased incidence of endometrial cancer. It was also known that factors such as obesity, which prolong exposure of the uterus to endogenous estrogens, resulted in an increased risk of developing endometrial cancer and possibly breast cancer.[5]

Clinical use of estrogen for women with premature surgical or natural menopause began in the 1930s. Clinical trials of estrogens from the urine of pregnant mares were initiated in 1941. In 1943 these preparations became available in the United States for use as oral postmenopausal estrogen therapy and were introduced onto the market in the United Kingdom in 1956. Use of postmenopausal estrogen therapy became widespread in the United States in the 1960s. Approximately 13 percent of women in the United States aged forty-five to sixty-four used postmenopausal estrogen therapy during that decade.[6]

In 1976 the epidemiologists R. Hoover, L. A. Gray Sr., Phillip Cole, and Brian McMahon published a prospective study of women given estrogens for menopause symptoms. The women were followed for twelve years for incidence of breast cancer, and the researchers reported a borderline statistically significant increase in the risk of breast cancer. Subsequently, many studies confirmed the increased risk of breast cancer associated with the use of postmenopausal estrogen supplement therapy, and data on the use of estrogen-progesterone oral contraceptives revealed a relatively consistent finding of an increased risk for breast cancer occurring before the age of forty-five and particularly before thirty-five.[7] In 1999 the IARC concluded that estrogens and combined estrogen-progesterone postmenopausal medication and oral contraceptives caused breast cancer. In addition, endometrial cancer was also found to be caused by postmenopausal estrogen supplement therapy.[8]

• • •

During World War II, the chemist Elwood Jensen studied chemical warfare agents at the University of Chicago, but when two hospitalizations resulted from exposure to the vigorous production of novel reactions, he decided to learn about steroid hormones. After the war, Jensen went to work with Charles Huggins, who would be awarded the Nobel Prize in 1966 for discovering that castration slowed the progression of prostate cancer. After developing a radioactive estrogen, Jensen was able to see its selective uptake in the reproductive tissues of the rat. Next, Jensen demonstrated in 1962 that the estrogen was binding to a receptor and that this binding could be inhibited by a chemical that inhibited the effect of estrogen.[9] This first receptor discovered would be called the alpha estrogen receptor, and others, including the beta receptor, would follow it. The alpha estrogen receptor was found to be located in the breast, uterus, hypothalamus, and ovary, whereas the beta receptor was found in the kidneys, prostate, heart, lungs, intestine, and bone.

Anna Soto and Carlos Sonnenschein, at the Tufts University School of Medicine in the 1980s, were investigating cell division in cultured human breast cancer cells. They found that these cells continually divided but that when human blood was added to the culture medium, proliferation stopped. Addition of estradiol, the main form of estrogen found in women, reversed this inhibition, and the cells continued to divide.[10] But all of a sudden, something went wrong; their cells no longer responded to the inhibitory factor in the human serum. After months of searching for the possible culprit, they discovered that it was a component of the plastic tubes that they were using in processing their cultures.[11] In 1991 Soto and her coworkers reported that the offending chemical was nonylphenol, a chemical released from polystyrene centrifuge tubes that was later found to interact with the estrogen receptors of the cultured human breast cancer cells. Nonylphenol had been added to the plastic because of its antioxidant properties. The estrogen effects of nonylphenol present in commercially available plastic centrifuge tubes were verified by observing estrogenic changes in the endometrium of rats.[12]

In the same year, Soto was one of twenty scientists from academia and government at the initial Wingspread conference in Racine, Wisconsin, organized by Theo Colborn, of the World Wildlife Fund. They published a consensus statement of this gathering and coined the term "hormone disruptors" to identify a wide range of biological effects. The consensus report pointed the hormone disruptor finger at several pesticides and other chemicals, PCBs, dioxin, cadmium, lead, mercury, soy products, and laboratory animal and pet food products. By the time of this meeting, Soto and her coinvestigators had standardized their testing protocol further and called it the E-SCREEN. Nonylphenol had the greatest potency of the tested industrial chemicals, but it was still only 1/100,000th as potent as estradiol. The pesticides DDT and kepone were even less potent, at one millionth that of estradiol, and others such as PCBs and chlordane were even less potent. The only chemicals tested that approached the potency of estradiol were naturally occurring and derived from fungi that contaminate certain grains, which had potencies about one hundredth that of estradiol.[13] By 1995 Soto and her colleagues had added other chemicals to their list of positive results in the E-SCREEN: the pesticides dieldrin, endosulfan, and toxaphene and other plastic components such as phthalates and bisphenol A.[14]

The next year, Theo Colborn published the book *Our Stolen Future*, which attempted to scare the public into believing that small amounts of chemicals in the environment with "estrogenic" and other hormonal effects were having a widespread effect on wildlife and humans. As opposed to Rachel Carson, who made a convincing argument about indiscriminate pesticide use poisoning wildlife, Colborn's arguments were largely theoretical and based on Soto's E-SCREEN experiments. Colborn's claims regarding pesticides and PCBs causing cancers such as breast cancer caused an immediate swell of outcry for studies.[15] But other scientists were not convinced. A formidable group of scientists from academia and the National Research Council of the National Academy of Sciences published a comprehensive review of the current literature and found little scientific evidence for hormone disruption caused by levels of industrial chemicals in the environment.[16] Robert Nilsson, of the Swedish National Chemicals Inspectorate, also took exception to the concept that weak estrogens could cause effects when there are so many powerful estrogens involved naturally in a woman's body.[17]

When Soto's reports suggested a link between PCBs and breast cancer, researchers came under considerable pressure to evaluate the possible connection. Two small epidemiology studies had been published, one in 1984 and the other in 1992, with contradictory results of associations of PCBs in their body fat with breast cancer. Available animal studies provided conflicting data as well: PCBs were shown to increase liver tumors, but the number of mammary tumors in female rats were not increased and possibly decreased. The first larger study of 150 women with breast cancer matched to 150 women controls looked at PCB and DDT levels in blood that had been drawn at the time of diagnosis. This study by Nancy Krieger, from the Kaiser Foundation, and Mary Wolff, of Mount Sinai in New York, was reported in 1994 and found no association with either chemical. The next study from Harvard of 240 women was published in 1997 and even found a protective effect for high levels of PCBs and DDE, the major metabolite of DDT.

Precautionary politics brushed these results aside, resulting in calls for more research on PCBs, when in the 1990s a highly publicized greater rate of breast cancer on Long Island was found compared to the rest of New York State. Eventually about twenty large and expensive studies were done, including one specifically looking at Long Island women with breast cancer. Their findings still did not show that PCBs caused breast cancer, and scientists at the American Cancer Society in 2002 concluded that "the evidence does not support an association. This conclusion applies particularly to the study of DDT, DDE, and to all PCBs combined."[18]

Additionally, animal studies reported in 1998 showed that PCBs were protective against the development of mammary tumors. A large bioassay paradoxically showed that although animals dosed with PCBs got liver cancer, they lived longer. How could that happen? The female rats used in the bioassay had a high spontaneous incidence of lethal mammary tumors, and the PCB-dosed animals had much less mammary cancer than did the control animals, so they lived longer. The most likely explanation, based on pharmacology principles, is that even though PCBs were shown to cause estrogen-like effects in Anna Soto's E-SCREEN, they were actually weak estrogens compared to the estrogens circulating in the rat's bodies. Since they bound to the same estrogen receptor and had a weaker effect, the PCBs effectively acted like some of the antiestrogens used to prevent breast cancer in high-risk women.

Another chemical found to cause mammary cancer in rodents is atrazine. Atrazine is an herbicide used for weed control on most corn, sorghum, and sugarcane acreage. As a consequence, atrazine is found as a water contaminant in rivers, lakes, and groundwater. Atrazine causes mammary cancer in rats, which prompted the EPA to consider banning or severely restricting its use. As in any mammal, mammary cancer in the rat is linked to estrogen; increased levels of estrogen cause proliferation of mammary tissue. The Sprague-Dawley strain of rat used for the atrazine study was the same as that used in the PCB bioassay, and this strain has a notably high spontaneous rate of mammary tumors. Unlike PCBs, atrazine exposure increased this already high rate of mammary tumors.

The reason the mammary cancer rate is so high in these rats is that when they go into reproductive senescence, similar to menopause in women, their ovaries increase production of estrogen. In contrast, most mammals, including humans, produce much less estrogen after menopause. Atrazine affects the hypothalamus and causes these rats to go into menopause sooner than normal, thereby signaling the pituitary gland to increase ovarian estrogen production. As a consequence, atrazine causes this strain of rats to receive a much higher level of estrogen over a longer period of time, causing mammary tumor development. This effect does not happen in other strains of rats, which also did not show increases in tumors with atrazine exposure. However, it stands to reason that atrazine would not cause breast cancer in humans, since in women menopause brings lower estrogen levels, in contrast to the higher levels produced in this particular strain of rats.[19]

To this day, we still don't know whether *any* industrial chemical can cause breast cancer in humans, even though many cause mammary tumors in rodents. The National Toxicology Program has found forty-eight chemicals that cause increases in tumors of the mammary glands in animals, but none have been found to cause breast tumors in women. The only chemicals known to be associated with an increased risk of breast cancer in women are still the hormonal preparations for birth control or replacement therapy.

During the 1940s and 1950s in Australia, another type of problem was linked to estrogens of a very different sort. Sexually mature sheep were infertile and suffered mammary enlargements, and their offspring exhibited several types of developmental abnormalities. Alterations were found in vaginal, uterine, and cervical development, along with alterations in ovulatory frequency and premature reproductive senescence. The problem was that the clover the sheep were eating contained several estrogenic chemicals called phytoestrogens, which are naturally occurring estrogenic compounds that protect the plants against fungi.[20]

Humans are also exposed to estrogens in our food. Approximately three hundred plants produce estrogen, and the amount varies with the plant's growing conditions. Phytoestrogens come in multiple forms; coumestans, whose structure mimics estrogen, are present in plant material. Soybeans also contain isoflavones such as genistein, which is a major dietary phytoestrogen also found in chickpeas and peanuts. It appears that some of the most potent phytoestrogens inhibit endogenous estrogen responses in humans in ways similar to PCBs; therefore, they do not appear to be a risk for breast cancer.[21]

• • •

In Colborn's book, pesticides and PCBs were reported to cause detrimental effects on men's reproductive function. She stated that "the most dramatic and troubling sign that hormone disruptors may already have taken a major toll comes from reports that human male sperm counts have plummeted over the past half century, a blink of the eye in the history of the human species." However, when the Committee on Hormonally Active Agents in the Environment of the National Research Council of the National Academy of Sciences reviewed these data, it was not clear to them that sperm counts were actually declining. Apparently, the data from the various studies used in the study Colborn was relying on had a number of flaws.[22] Some reviews have come to the same conclusion,[23] and others have not.[24]

In contrast to the suspect data, researchers at the Danish Ramazzini Centre of Aarhus University Hospital recruited a Danish pregnancy cohort in 1988–1989 and studied their male offspring. While these women were pregnant, maternal serum concentrations of PCBs and DDE were determined. Then, sperm concentration; total sperm count, motility, and morphology; and reproductive hormone levels were measured in their male offspring twenty years later. This study did not find any correlations between levels of PCBs or DDE of mothers and altered sperm or reproductive hormone measures in the sons.[25]

The hypothesis predicted that prenatal exposure of male fetuses to maternal levels of estrogenic pesticides and PCBs were the culprits. However, there are reasons to question whether small levels of these weak estrogenic chemicals could alter male fetal sexual development. The first of these is similar to the case for breast cancer risk from weakly estrogenic chemicals, that is, the large amount of competing estrogens in the maternal circulation and fetus, which would overwhelm any estrogenic effects of PCBs.

Second is the observation that even the powerful estrogen DES given to women resulted in relatively small effects on the sperm counts of their male offspring: some studies showed no effects, some showed increases, and some showed decreases. In contrast to these findings in humans, studies of in utero exposures of DES in mice showed decreased sperm production and sperm abnormalities.[26] But, one possible explanation for this difference is that the endogenous levels of estrogens in pregnant humans are about a hundred times greater than in mice.[27] Christopher Borgert, of the University of Florida's Department of Physiological Sciences, showed that thresholds for effects of DES in fetal male rats are lower than for humans, refuting the notion that rodents are less sensitive—they are more sensitive to the anti-androgenic effects of DES.[28] Again, here is an example of a disconnect between studies in rodents and in humans.

Chemicals in the environment that mimic or inhibit estrogen effects have received the most attention. But there are also chemicals that affect other glandular functions, such as those of the thyroid, adrenal, and pituitary glands. Like many of the glands in the body, the thyroid gland operates under stimulation from the pituitary gland via thyroid stimulating hormone (TSH). In clinical medicine, the lack of thyroid hormone is most easily detected by an elevation in TSH through relaxation of the negative feedback mechanism of thyroid hormone on the pituitary. The presence of too much thyroid hormone signals the thyroid to produce less TSH, and a lack of thyroid hormone stimulates more TSH production by the pituitary.[29] Some chemicals, including drugs specifically designed to do so, can cause a decrease in thyroid hormone production. In that event, a growth signal on the thyroid-stimulating hormone (TSH) is delivered that causes more thyroid cells to be produced by the division of existing cells. If the delicate feedback mechanism goes haywire and too much TSH is called for, the excessive cell division increases the odds of tumor formation in rodents. Humans, on the other hand, appear to be resistant to this particular mechanism of thyroid tumor formation. Even people with very large thyroid glands (goiters) do not develop tumors in this way.

For decades, sulfa drugs have been the gold standard for the treatment of urinary bladder infections, and two of the most widely used drugs are sulfamethazine and sulfamethoxazole. They produce thyroid tumors in rodent bioassays by interfering with thyroid hormone production, stimu-lating TSH production by the pituitary, increasing cell proliferation in the thyroid, and causing growth of the thyroid gland and tumors. Not only are humans much less prone to these tumorigenic effects of thyroid hormone imbalance; they also are much less sensitive to inhibition of the thyroid hor-mone production by the sulfa drugs than rodents. Thyroid hormones are produced when an enzyme called thyroperoxidase attaches iodine atoms to the amino acid tyrosine. Sulfa drugs inhibit both the human and rodent form of this enzyme, but it takes one thousand times the concentration to inhibit the human enzyme compared to the rodent version. So, because we have a different peroxidase enzyme and are less sensitive to any resulting thyroid growth, humans can treat their bladder infections without getting thyroid cancer.[30]

There are many other chemicals that can be demonstrated either to act like hormones or inhibit them. Research in animals and humans have and will continue to identify others besides those discussed here. But again, the truth about these chemicals is not as obvious as it may seem at first. Differ-ent findings between rodents and humans for hormone mimics provide fur-ther evidence for using caution when extrapolating chemical effects from one species to another. And as for hormone mimics, the illustration of the findings of the human breast cancer cell E-SCREEN test compared to the lack of estrogenic effects in humans gives rise to concerns about in vitro (test-tube-based) screening methods in general. In the next chapter we will explore how to provide better tests for human disease potential for chemi-cals so that we can move beyond the standard rodent bioassays. Many of these involve in vitro experiments that can be done relatively rapidly and in great quantities.

27 | Building Better Tools for Testing

I n a philosophical treatise of fundamental methods used in science, Irving Copi, at the University of Michigan, stated that "older theories are not so much abandoned as corrected" and that "ultimately, our last court of appeal in deciding between rival hypotheses is experience." So it should be with the results of the rodent bioassay and in vitro tests used for predicting human cancer and other diseases. As we have seen earlier, there is much we don't know about the validity of the rodent model for human disease. Therefore, the use of animals that are more closely related to humans, such as primates, would seem to solve most of the problems of genetic differences with humans. Monkeys have, in fact, been used productively to study many chemicals that had been previously found to produce tumors in rodents. Saccharin, for example, was given to monkeys for up to twenty-four years without ill effect, and this information provided additional assurance that the bladder tumors found in rats caused by saccharin wouldn't happen in humans.[1] But the very kinship with humans that makes primates valuable for research is what prohibits their widespread use. The difficulty and cost of keeping primates in a humane environment for extended periods, as well as the ethical considerations of using them as test subjects, preclude them from being used on a routine basis. Imagine testing one thousand monkeys per chemical bioassay!

Another alternative to the default rodent bioassay would be to use a more tailored approach for testing, that is, to choose the best-suited animal species for each type of exposure and predicted response. Before the bioassay testing programs settled on rats and mice, many different species were used, including dogs, pigs, hamsters, and guinea pigs. Fish have been selected for some tests, but they obviously can't be used for inhalation studies, which is the route of occupational exposure for most industrial chemicals. Guinea pigs and hamsters exhibit similar responses to humans for certain chemicals but not for others. Each of these species has had some amount of success as a laboratory subject, but none has been advantageous enough in terms of cost effectiveness and historical precedent to keep government agencies from settling on rats and mice for most types of testing.

Efforts have been made to improve the speed and relevance of the rodent bioassay by engineering mice whose genes have defects similar to those exhibited by human cancer. These genetic alterations are designed to enhance cancer susceptibility in the animals within the assay period, which could be shortened to six to twelve months instead of two years. The concept is that exposure of these genetically modified animals to potentially carcinogenic agents would result in a rapid, less costly induction of tumors more relevant to humans. The technology is called transgenic modification, whereby embryonic cells are injected with modified genetic material that becomes incorporated into the host genome. Transgenic mice have been used to study many human diseases, especially those caused by genetic mutations. In the case of the cancer bioassay, the genetic changes in the mice are like those chemically initiated by a genotoxic carcinogen. This is similar to the initiation-promotion method developed in the 1930s by Berenblum and Shubik, whereby genotoxic chemicals initiate the cancer process and then the animal is exposed to the test chemical. However, transgenics allow the investigator to control the specific mutation desired.

Four types of transgenic mice have received the most attention: the Tg .AC, rasH2, $p53^{+/-}$, and $XPA^{-/-}p53^{+/-}$, two of these having defects in the *p53* tumor suppressor gene, which is the one that is mutated in about 50 percent of all human cancers. Transgenics have been a boon to researchers because they have decreased the time to tumor development, reducing the cost of the standard chemical-induced bioassay. Unfortunately, it is still not clear whether results obtained with transgenic mice are more or less meaningful than the standard assay for human cancer prediction. One obvious problem with these mice is that the alterations that went into making them don't necessarily mimic the susceptibility of humans to cancer processes caused by chemicals. For example, although many human tumors do have the *p53* mutation, it is not usually the first event in the cancer process.

An evaluation in 2013 showed that among the tested agents identified as human carcinogens or likely human carcinogens, 86 percent, 67 percent, and 43 percent were detected as carcinogenic in the rasH2, the Tg.AC, and the Trp53$^{+/-}$ models, respectively. The overall conclusion is that these assays appear to be moderately efficient in detecting human carcinogens and very effective in correctly identifying chemicals without human carcinogenic potential.[2] One problem with the use of the transgenic models is dose-response assessment, which is particularly problematic since these models are, at times, much more or less sensitive than the conventional rodent cancer bioassays. However, the biggest problem with transgenics is that they again leave us with lots of positive animal bioassay results for which we don't have corresponding human studies, thereby increasing the amount of information that is uncertain for human cancer prediction.[3]

• • •

There have been major efforts to develop alternative approaches to identifying carcinogenic chemicals. From a hypothetical perspective, tumors produced in experimental animals by genotoxic mechanisms are viewed as potentially relevant to humans because of a presumed common underlying mechanism: we all have DNA and mostly similar chromosomes, which are vulnerable to attack. Yet subtle differences in genomics between species can add up to substantial differences in cellular functions. Since most human carcinogens involve genotoxic mechanisms, an argument can be made that the further development of reliable tests for genotoxicity could yield better predictors of human carcinogenicity than the much more time-consuming and expensive animal bioassays.

As we learned in our discussion of cancer mechanisms, the well-known University of California–Berkeley biochemical geneticist Bruce Ames invented a bacterial test for mutagenicity. At first, this test was used to understand the mechanism by which certain mutagens cause cancer. However, the Ames test became the first "high-throughput" test for screening potential human carcinogens because bacterial colonies grow quickly. Studies before 1978 using this test showed a high degree of association with cancer bioassay results because most of these involved genotoxic carcinogens. Ames and his fellow scientists assessed the studies of three hundred chemicals up to 1975 in their test. There was 90 percent accuracy, according to them, for their tests to predict rodent carcinogens, given that 156 of 174 carcinogens were identified.[4] In addition, an analysis by scientists at Imperial Chemical Industries, in the United Kingdom, found that the Ames test was accurate in predicting 94 percent of carcinogens. The chemicals tested were heavily weighted toward suspected human carcinogens, such as PAHs found in smoke, aromatic amines from dyestuffs, and chemotherapeutic alkylating agents.[5]

However, the value of the Ames test for predicting animal carcinogens testing over the past four decades has become increasingly limited. It doesn't detect most animal carcinogens, since those that have been tested in recent years are mostly not genotoxic. Earl Zeiger, from the National Institute of Environmental Health Sciences (NIEHS), summarized the state of genotoxicity testing from 1975 to 1998. Zeiger noted that later databases did not show as good a correlation as more and more chemicals were tested. This is largely attributable to the attenuation of the results for tested chemicals by rodent carcinogens that cause tumors by way of high-dose toxicity and other effects rather than by direct alteration of the genome, as is the case for most human carcinogens.[6] Ames tests come up negative for most of these, as might be expected, so his tests are not optimal predictors of the results of animal bioassays when the carcinogens in question are not genotoxic—which is only about 20 percent of the time.[7] But by some baffling perversion of logic, the basis on which genotoxicity tests have usually been evaluated is whether they can predict animal carcinogens as opposed to human carcinogens. This is unfortunate because relatively easy-to-perform mutagenicity tests appear to be more valid predictors of human cancer risk.

In attempts to improve prediction, hundreds of other genotoxicity tests have been developed, and several are in use today. Yeast, molds, fruit flies, plant cells, various bacteria besides Salmonella, and a wide variety of mammalian cell types from mice and rats, Chinese hamsters, and humans have been used. They have enabled us to study various aspects of cancer biology, such as mutations, changes in chromosome structure, transfers of segments between chromosomes, and increased DNA repair, but there is no consensus on how to apply the results to human cancer risk. Moreover, correlations between these tests and cancer findings in rodent bioassays are generally no better than the Ames test, although the combination of this test with two other tests improves predictability by some measures.[8]

Gary Williams, who was my division director at the American Health Foundation, was one of the leaders in the field of genotoxicity testing and developed the "Williams test." This was a method of looking for DNA adduct formation by detecting the repair of this damage by cells. The types of cells used were hepatocytes, so there was no need to add any metabolic activation; the cells were capable of forming the active metabolites through their cytochrome P450 enzymes. Gary and John Weisburger proposed a "systematic decision point approach for detecting and evaluating substances for carcinogenic risk." They proposed that the initial evaluation of a chemical would include a battery of five short-term in vitro tests including the Ames and Williams tests. If clear-cut evidence of genotoxicity in more than one test is obtained, the chemical is highly suspect of being a human carcinogen.[9]

The remainder of carcinogens act through nongenotoxic receptor-mediated gene expression, immunosuppression, or toxic irritations leading to cell proliferation. Nongenotoxic cancer mechanisms that involve biochemical alterations or chemical effects on gene expression have often been found to be species specific, meaning that animal studies involving those mechanisms are not relevant to humans. Other times, the tumors are caused by the very high doses of chemicals the animals are subjected to, producing a chronic toxicity and irritation that wouldn't happen at typical human dose levels; therefore, these studies are not relevant to humans either. These are the types of mechanisms that cause the majority of the positive rodent bioassays and comprise what the former head of the IARC Monograph Program Lorenzo Tomatis called the "parking lot": chemicals possibly carcinogenic to humans. The question for the future of cancer identification is whether there is an alternative to putting more chemicals in the "parking lot" of animal carcinogens with unknown relevance to human cancer.

• • •

Not addressed by standard bioassays or transgenics is the need to test tens of thousands of old and new chemicals that have no or little toxicology information available. The proposed answer is called "high-throughput" testing, similar to genotoxicity tests, that can be done in vitro with little expense and rapid results. The precursor to this testing method for toxicity was high-throughput screening for efficacy, widely used as a drug-discovery process in the pharmaceutical industry. It leverages automation to quickly assay the pharmacological activity of a large number of compounds.

For use in toxicology, the key to these high-throughput methods lies in determining chemical-induced changes in cellular-response networks that result in toxic effects. These are interconnected pathways composed of interactions of genes, proteins, and small molecules that maintain normal cellular function, control communication between cells, and allow cells to adapt to changes in their environment. Pathways that can lead to adverse health effects when sufficiently perturbed by chemicals are termed toxicity pathways. One can think of these high-throughput tests as cousins of the Ames test for mutagenicity, where the cellular-response network involves changes in DNA that affect key genes involved in cell cycle regulation.[10]

Sounds simple? Well the devil is in the details. Some high-throughput tests involve the use of established human cell lines in culture, many of which can be readily purchased. However, their use can be problematic because sometimes they are not the expected type of cells, since the cell cultures have been contaminated by different cell lines, which can overcome the original cells. This was revealed by Christopher Korch, at the University of Colorado, who calls himself "a corrector."[11] Korch and his colleagues identified 574 articles between 2000 and 2014 that provided an incorrect attribution and only fifty-seven articles that provided a correct attribution for HeLa cells, an immortal cervical adenocarcinoma cell line.[12] As a consequence of their research on numerous cell lines, thousands of scientific articles in a thousand journals have been published about the wrong cells. The amount of money spent on this flawed research was estimated to be $713 million—and the amount of money wrongly spent on subsequent work based on these errors was estimated to be $3.5 billion. The impact of these cell line errors on high-throughput testing has not been reported, but it could be substantial.[13]

An example of a relatively straightforward high-throughput test is the identification of thyroid peroxidase inhibitors, which can cause hypothyroidism. In one study, thyroid peroxidase from rats was used to screen inhibitors of the enzyme using high concentrations of over a thousand chemicals on special assay plates using automated laboratory procedures. The most potent thyroid peroxidase inhibitors in this assay are drugs used to treat hyperthyroidism, which makes sense.[14] To determine the concentration of these inhibitors, which bind to thyroid peroxidase at the cellular level, another type of computer modeling is done to estimate the amount of a chemical that is absorbed into the bloodstream, distributed throughout the body, and eliminated. The combination of these two models can be used to estimate the inhibition of thyroid peroxidase in a person caused by a level of exposure to inhibitors in the environment.[15]

But thyroid peroxidase inhibition is a relatively simple example compared to the ambitions of determining several interacting chemical-induced changes in cellular-response networks necessary to produce cancer or some other diseases. First, we have to identify the significant networks, their components, and the direction and magnitude of these changes. Then, the components have to be measured accurately. Finally, the results of testing need to be verified against the toxic endpoint. This type of testing requires that we know the mechanistic steps that lead to a toxic response, in contrast to the "black-box" approach of a bioassay in which we see the outcome of a chemical exposure but not the mechanism(s) by which it occurs. Similar to the results of transgenic mouse testing, the predictions of high-throughput testing need to be verified for human disease and not just for disease in experimental animals. So again, we have the dilemma of having so much information gained for diseases, including cancer, in experimental animals and a lack of correlation in many instances with human disease.

Even so, these high-throughput assays are continuing to be developed for many toxicity and cancer endpoints. They probably represent the best response for the need to test the thousands of chemicals for which little is known toxicologically. However, we need to be wary of the "streetlight effect," the observational bias whereby people only search for something where it is easiest to look.

28 | An Ounce of Prevention Is Worth a Pound of Cure

What does all of this information about chemicals, toxins, and cancer mean? In the introduction, I proposed that the history of toxicology shows that we have made great progress understanding diseases caused by chemicals and toxins. By 1960 we had harvested the low-hanging fruit on the chemical causes of disease. We have mostly picked the tree clean since then and now need to turn our attention to the contents of the harvest. To follow in the footsteps of Ernst Wynder and apply our knowledge for the public good, our attention should be focused on "what we eat, what we drink, and what we smoke." Given our more recent understanding of threats to our health, I would add "what we burn for energy, what we use for recreational drugs, and how we exercise." Wynder used epidemiology to identify diseases caused by lifestyles and chemicals and then performed toxicological studies in the laboratory to understand chemical-induced disease processes and how to prevent them. Rachel Carson made a strong argument in *Silent Spring* that cancer prevention is just as important and relevant as potential cures for cancer.[1] She likened this approach to that for infectious disease control, in which both prevention and cure are given equal attention.[2]

Why focus on prevention rather than solely on therapies for diseases such as cancer? The short answer is that prevention is much more effective in terms of both health and cost. We haven't identified the cause of many cancers, but where we have identified likely causes, prevention should be the path of least resistance forward. We understand where these efforts could be focused to derive the biggest public health benefit, but reducing some well-understood causes of cancer requires personal and public willpower. I would argue that cigarette smoking, excessive alcohol use, drug abuse, overeating, air pollution, and lack of exercise appear to be the primary preventable causes of cancer and other fatal diseases that have been identified.

What were some of these low-hanging fruits of chemically induced disease that had been identified by 1960? There were early notable successes in the prevention of occupational cancer. The diseases of miners such as lung cancer, silicosis, and asbestosis were identified decades before 1960. After World War II, studies of workers in shipyards and other occupational settings identified asbestos as a cause of lung cancer and asbestosis. Cancers and other diseases produced by chemical industries, dyes, metal smelting, beryllium, lead (in occupationally exposed adults), mercury, and phosphorus were also identified before 1960 and largely eliminated. Even after 1960, cancer from occupational exposure to benzene, vinyl chloride, and other chemicals have been largely identified and controlled.

However, for carcinogen identification, the most important example was the association between cigarette smoking and lung cancer. Although Wynder and Doll would eventually estimate that approximately one-third of all cancers resulted from cigarette smoking, their findings were largely ignored by the general public and media for decades. Still, smoking tobacco remains a leading cause of preventable disease and death in the United States, with approximately 480,000 premature deaths and more than $300 billion in direct health care expenditures and productivity losses each year.[3] According to the Centers for Disease Control, in fiscal year 2016, states will collect $25.8 billion from tobacco taxes and legal settlements, but shamefully these states will only spend $468 million—less than 2 percent—on prevention and smoking cessation programs.[4] Just in economic terms, this expenditure on smoking cessation is about one-thousandth of the health costs of smoking. Unfortunately, the rest of the tobacco settlement money usually goes to fill state budget deficits. State and local governments should change this short-sighted policy to prevent smoking-related cancers and other diseases.

Prevention of some chemical exposures has affected the lives of children and their productivity after they become adults. Certainly, the prevention of lead poisoning among most children has been an important success story; however, the problem still exists today. A study of children six years old or younger published in 2009 found that the remaining health care costs for children with elevated lead levels was as high as $50 billion. If we add in the loss of earnings, tax revenue, special education, and direct costs of crime related to the long-term effects of lead, the price goes up to as much as $269 billion.[5]

In the chapter on toxicology and climate change, we learned that health effects from burning fossil fuels result in costs adding up to around $500 billion per year. According to the National Institute on Drug Abuse, alcohol and illicit drugs cost our nation more than $400 billion annually from crime, lost work productivity, and health care.[6] Those few categories—cigarette smoking, lead poisoning, fossil fuel use, and drug abuse—are still costing over a trillion dollars per year.

For some obvious public health concerns, there has been notable success. For example, the incidence of cigarette smoking in the United States has precipitously decreased, from over 42 percent of the adult population in 1965 to 17 percent in 2014.[7] Why did this happen? Setting an example, the smoking rate among physicians declined from 60 percent in 1949 to 30 percent by 1964. During the decade between 1950 and 1960, statements were issued by a number of organizations that called attention to smoking as an important health hazard, particularly with respect to lung cancer and cardiovascular disease. These organizations included the American Cancer Society and the American Heart Association as well as the British Medical Research Council; the cancer societies of Denmark, Norway, Sweden, Finland, and the Netherlands; the Joint Tuberculosis Council of Great Britain; and the Canadian National Department of Health and Welfare.[8] The 1964 publication of "Smoking and Health: Report of the Advisory Committee to the Surgeon General" was a starting point for concerted tobacco-control efforts in the United States. However, the government took the position that it was not its place to tell people to stop smoking, and the wording of the label, "Caution: Cigarette Smoking May Be Hazardous to Your Health," contrasted sharply with the certainty of the 1964 report's conclusion on smoking and lung cancer. Although many individual physicians rapidly accepted the smoking and health findings, the American Medical Association (AMA) took more than two decades to take a clear stand on the issue. Next, the Public Health Cigarette Smoking Act of 1969 prohibited broadcast advertising of cigarettes, after which the tobacco companies responded by switching media. The five major tobacco companies spent $62 million on magazine advertising in 1970, the year before the broadcasting ban was implemented, but by 1976 they were spending $152 million.[9]

In the early 1970s a movement emerged to promote the interests of nonsmokers and those who wanted to quit. Surgeon General C. Everett Koop during the Reagan administration advocated for smoke-free public places to a greater degree than any of his predecessors, and federal and state taxes on cigarettes began to climb as part of the ripple effect of the changing public discourse. During the 1990s stores begin selling over-the-counter nicotine medications, and the National Cancer Institute conducted a large nationwide intervention study: the "American Stop Smoking Intervention." Also during this period, class-action litigation and litigation on behalf of state governments allowed plaintiffs to demand the release of internal documents of tobacco companies, and the resulting settlements led to large cash awards.[10] Finally, in 2009 the landmark passage of the Family Smoking Prevention and Tobacco Control Act (FSPTCA) granted the Food and Drug Administration the power to regulate tobacco products.[11] The progress in controlling cigarette smoking to prevent this major cause of disease has been slow but steady. Unfortunately, many lives were lost, and still more will die of these diseases in the future.

• • •

Perhaps the biggest barrier to prevention is the medical emphasis on treatment. Treatment of disease is taught in medical school; prevention is relegated to schools of public health. There are about 180 medical schools and about one-third that number of schools of public health, although schools other than medical schools have public health programs as well. Monetary support for research emphasizes treatment rather than prevention. Successfully treated patients give money to hospitals and medical schools, but whom do we thank for *preventing* a potential disease?

The difficulties of treatment, including hospitalizations, are well known to most of us personally. Siddhartha Mukherjee's book *The Emperor of All Maladies* explores the history of the benefits and risks of cancer treatments. There is no lack of "treatment" options for most cancers, but there are few highly successful treatments that involve minimal suffering. A recent study found that hospital medical error is the third most prevalent cause of death after cancer and cardiovascular disease.[12] While most treatment options involve hospitalization, preventative options for patients rarely do. Even without hospital error, the collateral damage of surgery and chemotherapy makes the calculation for prevention compelling.

But there are many barriers to prevention of disease. If doctors routinely addressed exposures that are under individual control, the first requirement would be to educate patients to appreciate dangerous behaviors that could be prevented. Second, a person needs to understand the nature of the exposures and how to avoid them. Unfortunately, people often get mixed messages about the dangers of exposures. We have seen how the tobacco companies either denied or sowed doubt about cigarette smoking and lung cancer. I remember my father repeating the propaganda of the tobacco companies in doubting the information about the dangers of cigarettes; he died in 1959 from lung cancer. More recently the book and film *Merchants of Doubt* showed how the climate change deniers use the playbook of the tobacco companies to sow doubt on the evidence for human-caused global warming.

Another barrier to prevention is a lack of self-motivation. For example, how does one get motivated to quit smoking? When I was thirty-three, I was having lunch with a friend right after Christmas and was suffering from a terrible sinus infection, sore throat, and cough. I knew that smoking could be deadly, and although I had tried to quit several times, I was addicted. My friend and I used to play high-stakes poker, and we were very competitive. So I proposed that we bet a thousand dollars and quit smoking on

New Year's Day; whoever started back first would lose the bet. My friend said that was a little too pricey for him but was willing to wager two hundred dollars if we quit right then and there. That was the end of my cigarette smoking. The key for me was motivation. It was peer pressure that started me smoking and peer pressure that ended it. So I believe that ways to increase the level of peer pressure to prevent smoking are most effective. School educational programs aimed at making smokers shunned by their peers are effective. The same approach is used and can be effective for drugs.

Often it is the unintended consequences of the well-motivated polemics that raise barriers to prevention because of the lack of alternatives. Addiction, whether to opiates or nicotine, is real and should be treated as a medical condition. There are alternatives to cigarettes for those smokers who are addicted to nicotine. One of these is the electronic cigarette, a pure nicotine delivery system without the tar. It has been estimated to eliminate 95 percent of the harm by the Royal College of Physicians, in the United Kingdom. But some doctors and scientists emphasize the dangers of e-cigarettes and focus on that remaining 5 percent or fear that children might get addicted. On top of that, the makers of e-cigarettes are not allowed to tell the public about the health benefits of switching because government regulation forbids it. Laws banning smoking in buildings may also apply to e-cigarettes, thereby providing an additional barrier to their use. This collective approach to e-cigarettes "is the same as asking, 'What are the relative risks of jumping out a fourth story window versus taking the stairs?'" said David Sweanor, a lawyer with the Center for Health Law, Policy, and Ethics at the University of Ottawa. "These guys are saying: 'Look, these stairs, people could slip, they could get mugged. We just don't know yet.'"[13]

A similar argument can be made for a more traditional alternative to cigarettes: smokeless tobacco. The Scandinavian scientific community has a very different attitude toward the use of this product than do U.S. scientists, and the Scandinavian products have been designed to contain fewer carcinogens than their American counterparts. As a result, about one-quarter of all men in Sweden use smokeless tobacco on a daily basis. Compared to the many types of cancers caused by cigarette smoking, smokeless tobacco causes cancers of the oral cavity, esophagus, and pancreas but not the others.[14] The bottom line is that people would be much better off using smokeless tobacco; however, this is not a message that people will get in the United States.

Deborah Arnott, the chief executive of Action on Smoking and Health, sees the U.S. approach to smoking rooted in the Puritan aspects of our culture, where abstinence is valued above all else. One could equally say this about our approach to drug abuse. Whereas during the Nixon years the tide was turned in favor of understanding that many addicts need their opiate addiction maintained, the tide again receded in favor of cold-turkey withdrawal rather than long-term maintenance. When I was working for the Nixon drug office, the conventional wisdom was that one-third of people would try heroin and hate it because it made them sick, one-third of people could use it occasionally without much of a problem, and another third would become seriously addicted and have a very difficult time with abstinence. We believed that the underlying cause of the addiction in the susceptible people was ingrained and that they needed maintenance treatment before they could become abstinent. I think that it's also fair to say that some of these addicts need to be on maintenance for a very long time.

Another barrier to accurate disease prevention is the consequential uncertainty from the large numbers of "cancer-causing chemicals" identified in animal bioassays. Animal experiments are often interpreted by government regulators as yielding more certain results than human studies. Remember from a previous chapter that occupational exposures have identified fewer than eighty chemicals and industrial processes as carcinogenic, whereas animal experiments have identified hundreds of chemicals as carcinogens. Governments rely on the results of animal studies to determine policy decisions about chemicals, which paints a confusing picture for the public, considering that we really don't know if harmful effects will occur. This raises serious doubts and questions about where our prevention dollars should be spent. Unfortunately, early on, chemicals like DDT and PCBs, which grabbed the most attention from the press after publication of the results of animal studies, got the lion's share of prevention dollars— instead of those identified lifestyle causes like smoking and diet that, based on human studies, are considered to cause 60 to 70 percent of cancer. Lorenzo Tomatis, a former director of the IARC, criticized the emphasis on lifestyle factors, which, "to the detriment of information on the role of chemical pollutants, favored the uninterrupted production of agents with negative effects on health that remain hidden or secret or are deliberately underestimated." In other words, his contention was that even though people like Ernst Wynder and Sir Richard Doll were working against tremendous headwinds in the media, by focusing on individual lifestyle choices, they were deliberately deceiving the public about the industrial chemical causes of disease. Tomatis also charged that this approach "unduly amplified the individual's responsibility."[15]

Consequently, there is a tension between those who want to interpret the positive animal tests as indicative of causing human cancer and those who want to place more emphasis on proven human carcinogens. On the one hand, the reliance on positive animal tests could be seen as a more conservative, more protective approach. But people that question individual responsibility for lifestyle factors play into the hands of the "merchants of doubt" who were opposed to regulating cigarette smoking and are now lined up against climate change concerns.

However, Tomatis was correct in one respect. Pollution was underestimated as a cause of disease; however, it is air pollution caused primarily by the burning of fossil fuels, not industrial chemical pollution. It was not until 2014 that the IARC published a monograph on diesel exhaust and not until 2016 that it published a monograph on outdoor air pollution. This latter monograph primarily addressed air pollution from transportation, power generation, industrial activity, biomass burning, and domestic heating and cooking. The IARC working group found that there was sufficient evidence of carcinogenicity in humans for outdoor pollution and particulates related to lung cancer. In addition, there were associations found for outdoor pollution and risk of bladder cancer. These findings argue for a more inclusive approach to the causes of cancer and that we need to listen to different points of view in toxicology.

In closing, this book has explored the historical development of toxicology, the discipline that investigates how toxins and chemicals cause cancer and other diseases. As a result, we now have a rich understanding of the ailments that people suffer from in their workplace and environment. But we also understand that much of the suffering is caused by our own devices, including lifestyle and our need for energy production.

In 1996, at the twenty-fifth-anniversary Symposium of the American Health Foundation, Ernst Wynder gave the final lecture and said the following: " 'To help people die young, as late in life as possible' is the adage we have chosen to guide our efforts in disease prevention at the American Health Foundation. Attaining this goal implies that we die free of disease after a full and productive long life. This enables us even at an old age to earn incomes, pay taxes, and contribute to the well-being of our nation."

Notes

Introduction

1. Exodus 15:23; Deuteronomy 29:18, 32:32–33; Jeremiah 9:15, 23:15; Hosea 10:4; 2 Kings 4:39–40.
2. Alexandra Witze and Jeff Kanipe, *Island on Fire.* (New York: Pegasus, 2014).
3. Numbers 21:6; Deuteronomy 32:24; Psalms 58:4; Proverbs 23:32; Jeremiah 8:17; Revelation 9:5.
4. For the reader who wants more detailed information about earlier developments in toxicology, I recommend the series edited by Philip Wexler, History of Toxicology and Environmental Health Sciences.

1. Cancer Clusters: Truth Can Be Obscure

1. Adrienne Mayor, *Greek Fire, Poison Arrows, and Scorpion Bombs* (New York: Overlook Duckworth, 2009).
2. John L. Creech Jr. and M. Johnson, "Angiosarcoma of Liver in the Manufacture of Polyvinyl Chloride," *Journal of Occupational Medicine* 16 (1974): 150–51.
3. Clark Heath Jr., Henry H. Falk, and John Creech Jr., "Characteristics of Cases of Angiosarcoma of the Liver Among Vinyl Chloride Workers in the United States," *Annals of the New York Academy of Science* 246 (1975): 231–36.
4. John Whysner, Carson Conaway, Lynn Verna, and Gary Williams, "Vinyl Chloride Mechanistic Data and Risk Assessment: DNA Reactivity and Cross-Species Quantitative Risk Extrapolation," *Pharmacology and Therapeutics* 71 (1996): 7–28.
5. Robert Baan, Yaan Grosse, Kurt Straif, et al., "A Review of Human Carcinogens—Part F: Chemical Agents and Related Occupations," *Lancet Oncology* 10 (2009): 1143–44.

6. Branham v. Rohm and Haas et al., First Judicial District of Pennsylvania Court of Common Pleas, Philadelphia County Civil Trial Division, N. 3590, Thursday, October 21, 2010, 52.

7. Jonathan Harr, *A Civil Action* (New York: Vintage, 1996), 18–37.

8. Gerald S. Parker and Sharon L. Rosen, "Woburn Cancer Incidence and Environmental Hazards 1969–1978," Massachusetts Department of Public Health, January 23, 1981; Stauffer Chemical Company, "Woburn Environmental Studies, Phase I Report," April 1983.

9. Harr, *A Civil Action*, 90–91.

10. Harr, *A Civil Action*, 95–98.

11. S. Lagakos, B. Wessen, and M. Zelen, "An Analysis of Contaminated Well Water and Health Effects in Woburn, Massachusetts," *Journal of the American Statistical Association* 81 (1986): 583–96.

12. EPA, Record of Decision for Wells G and H, http://www.epa.gov/region1/superfund/sites/industriplex/16796.pdf.

13. EPA, "Trichloroethylene (TCE): Teach Chemical Summary," revised September 20, 2007, https://archive.epa.gov/region5/teach/web/pdf/tce_summary.pdf.

14. EPA, "Toxicological Review of Trichloroethylene in Support of Summary Information on the Integrated Risk Information System (IRIS)," CAS no. 79-01-6, 2011, EPA/635/R--09/011F, https://cfpub.epa.gov/ncea/iris/iris_documents/documents/toxreviews/0199tr/0199tr.pdf.

15. Carol S. Rubin, Adrianne K. Holmes, Martin G. Belson, et al., "Investigating Childhood Leukemia in Churchill County, Nevada," *Environmental Health Perspectives* 115 (2007): 151–57.

16. Craig Steinmaus, Meng Lu, Randall L. Todd, and Allan H. Smith, "Probability Estimates for the Unique Childhood Leukemia Cluster in Fallon, Nevada, and Risks Near Other U.S. Military Aviation Facilities," *Environmental Health Perspectives* 112 (2004): 766–71.

17. Rubin, Holmes, Belson, et al., "Investigating Childhood Leukemia."

18. D. Austin, S. Karp, R. Divorsky, and B. Henderson, "Excess Leukemia in Cohorts of Children Born Following Influenza Epidemics," *American Journal of Epidemiology* 101 (1975): 77–83.

19. Richard Doll, "The Seascale Cluster: A Probable Explanation," *British Journal of Cancer* 81 (1999): 3–5.

20. Leo Kinlen, "Childhood Leukemia, Military Aviation Facilities, and Population Mixing," *Environmental Health Perspectives* 112 (2004): A797–98; Leo Kinlen, "Childhood Leukaemia, Nuclear Sites, and Population Mixing," *British Journal of Cancer* 104 (2011): 12–18; A. Balkwill and F. Matthews, "Rural Population Mixing and Childhood Leukaemia: Effects of the North Sea Oil Industry in Scotland, Including the Area Near Dounreay Nuclear Site," *British Medical Journal* 20 (1993): 743–48.

21. Mel Greaves, "A Causal Mechanism for Childhood Acute Lymphoblastic Leukaemia," *Nature Reviews Cancer* 18 (2018): 471–84.

22. Michael Goodman, Joshua S. Naiman, Dina Goodman, and Judy S. LaKind, "Cancer Clusters in the USA: What Do the Last Twenty Years of State and Federal Investigations Tell Us?" *Critical Reviews of Toxicology* 42 (2012): 474–90.

23. Michael Coory, Rachael Wills, and Adrian Barnett, "Bayesian Versus Frequentist Statistical Inference for Investigating a One-Off Cancer Cluster Reported to a

Health Department," *MC Medical Research Methodology* 9 (2009): 30, doi:10.1186 /1471-2288-9-30.

24. Craig Steinmaus, Meng Lu, Randall L. Todd, and Allan H. Smith, "Probability Estimates for the Unique Childhood Leukemia Cluster in Fallon, Nevada, and Risks Near Other U.S. Military Aviation Facilities." *Environmental Health Perspectives* 12 (2004): 766–71.

25. Stephanie Warner and Timothy Aldrich, "The Status of Cancer Cluster Investigations Undertaken by State Health Departments," *American Journal of Public Health* 78 (1988): 306–7.

2. Death from Arsenic and Venoms: Truth Can Be Obvious

1. Allam Smith, Claudia Hopenhayn-Rich, Michael Bates, et al., "Cancer Risks from Arsenic in Drinking Water," *Environmental Health Perspectives* 97 (1992): 259–67.

2. Badal Mandal and Kazuo Suzuki, "Arsenic Round the World: A Review," *Talanta* 58 (2002): 201–35.

3. Joseph Graziano, personal communication.

4. Michael Hughes, Barbara Beck, Yu Chen, Ari Lewis, and David Thomas, "Arsenic Exposure and Toxicology: A Historical Perspective," *Toxicological Sciences* 123 (2011): 305–32.

5. Jose Borgoño, Patricia Vicent, Hernan Venturino, and Antonio Infante, "Arsenic in the Drinking Water of the City of Antofagasta," *Environmental Health Perspectives* 19, (1977): 103–5.

6. Wen-Pen Tseng, "Effects and Dose-Response Relationships of Skin Cancer and Blackfoot Disease with Arsenic," *Environmental Health Perspectives* 19 (1977): 109–19.

7. C. Chen, Y. Chuang, S. You, T. Lin, and H. Wu, "A Retrospective Study on Malignant Neoplasms of Bladder, Lung, and Liver in Blackfoot Disease Endemic Area in Taiwan," *British Journal of Cancer* 53 (1986): 399–405.

8. Alan Smith, Elena Lingas, and Mahfuzar Rahman, "Contamination of Drinking-Water by Arsenic in Bangladesh: A Public Health Emergency," *Bulletin of the World Health Organization* 78 (2000): 1093–103.

9. Smith, Lingas, and Rahman, "Contamination of Drinking-Water by Arsenic in Bangladesh."

10. Samuel Cohen, Lora Arnold, Barbera Beck, Ari Lewis, and Michal Eldan, "Evaluation of the Carcinogenicity of Inorganic Arsenic," *Critical Reviews in Toxicology* 43 (2013): 711–52.

11. Yao-Hua Law, "Stopping the Sting," *Science* 362 (2018): 631–35. L. Albeck-Ripka, "Australian Jellyfish Swarm Stings Thousands, Forcing Beach Closings," *New York Times*, January 7, 2019.

12. R. Theakston and H. Reid, "Development of Simple Standard Assay Procedures for the Characterization of Snake Venom," *Bulletin of the World Health Organization* 61 (1983): 949–56.

13. H. Khoo, R. Yuen, C. Poh, and C. Tan, "Biological Activities of *Synanceja horrida* (Stonefish) Venom," *Nature Toxins* 1 (1992): 54–60.

14. John Whysner and Paul Saunders, "Studies on the Venom of the Marine Snail *Conus californicus*," *Toxicon* 1 (1963): 113–22.

15. John Whysner and Paul Saunders, "Purification of the Lethal Fractions of the Venom of the Marine Snail *Conus californicus*," *Toxicon* 4 (1966): 177–81.

16. Jean-Pierre Changeux, Michiki Kasai, and Chen-Yuan Lee, "Use of a Snake Venom Toxin to Characterize the Cholinergic Receptor Protein," *Proceedings of the National Academy of Sciences* 67 (1970): 1241–47.

17. Douglas Fambrough and H. Hartzell, "Acetylcholine Receptors: Number and Distribution at Neuromuscular Junctions in Rat Diaphragm," *Science* 176 (1972): 189–91.

3. Paracelsus: The Alchemist at Work

1. Carl Jung, *Psychology and Alchemy*, in *The Collected Works of C. G. Jung*, 2nd ed., vol. 12, ed. Herbert Read, Michael Fordham, Gerhard Adler, and William McGuire (Princeton, NJ: Princeton University Press, 1953), 227–317.

2. Philip Ball, *The Devil's Doctor: Paracelsus and the World of Renaissance Magic and Science* (New York: Farrar, Straus and Giroux, 2006), 24.

3. Ball, *The Devil's Doctor*, 70.

4. Nigel Paneth, Ezra Susser, and Marvyn Susser, "Origins and Early Development of the Case-Control Study," in *A History of Epidemiological Methods and Concepts*, ed. Alfredo Morabia (Basel: Birkhauser Verlag, 2004), 294–95.

5. Ball, *The Devil's Doctor*, 52–93.

6. Ball, *The Devil's Doctor*, 164–71.

7. Jolande Jacobi, *Paracelsus: Selected Writings* (Princeton, NJ: Princeton University Press, 1951), xlii.

8. Ball, *The Devil's Doctor*, 93–100.

9. Ball, *The Devil's Doctor*, 93–100.

10. Jacobi, *Paracelsus*, 5.

11. Ball, *The Devil's Doctor*, 75–79.

12. Paracelsus, *Seven Defensiones. Four Treatises of Theophrastus von Hohenheim Called Paracelsus*, trans. C. Lilian Temkin (Baltimore, MD: Johns Hopkins University Press, 1941).

13. Jung, *Psychology and Alchemy*, 35.

14. Ball, *The Devil's Doctor*, 117–21.

15. Nicholis Goodrick-Clarke, *Paracelsus: Essential Readings* (Berkeley, CA: North Atlantic Books, 1999), 73.

16. Roy Porter, *The Greatest Benefit to Mankind: A Medical History of Humanity* (New York: Norton, 1997), 203.

17. Ball, *The Devil's Doctor*, 196.

18. Ball, *The Devil's Doctor*, 288.

19. Ball, *The Devil's Doctor*, 139–43.

20. Ball, *The Devil's Doctor*, 154.

21. Ball, *The Devil's Doctor*, 223–33.

22. Ball, *The Devil's Doctor*, 223–33.

23. Ball, *The Devil's Doctor*, 223–33.

24. Ball, *The Devil's Doctor*, 178.

25. Porter, *The Greatest Benefit to Mankind*, 203.

26. Jerome Nriagu, "Saturnine Drugs and Medicinal Exposure to Lead: An Historical Outline," in *Human Lead Exposure*, ed. Herbert L Needleman (Boca Raton, FL: CRC, 1992).

27. Porter, *The Greatest Benefit to Mankind*, 181.
28. W. H. Brock, *The Norton History of Chemistry* (New York: Norton, 1992), 45–46.
29. Michael Gallo and John Doull, "History and Scope of Toxicology: Chapter 1," in *Casarett and Doull's Toxicology*, 4th ed. (New York: Pergamon Press, 1991), 3–11.
30. Ball, *The Devil's Doctor*, 236.
31. Jacobi, *Paracelsus*, 93.
32. Gallo and Doull, "History and Scope of Toxicology," 3–11.
33. Paracelsus, *Seven Defensiones*.
34. Goodricke-Clark, *Paracelsus: Essential Readings*, 29–30.
35. Ball, *The Devil's Doctor*, 164.
36. Porter, *The Greatest Benefit to Mankind*, 205–10.

4. Mining and the Beginnings of Occupational Medicine

1. Mark Aldrich, "History of Workplace Safety in the United States," *EH.net*, Economic History Association, https://eh.net/encyclopedia/history-of-workplace-safety-in-the-united-states-1880-1970-2/#5.
2. Milton Lessler, "Lead and Lead Poisoning from Antiquity to Modern Times," *Ohio Journal of Science* 88 (1988): 78–84.
3. Philip Ball, *The Devil's Doctor: Paracelsus and the World of Renaissance Magic and Science* (New York: Farrar, Straus and Giroux, 2006), 70.
4. "Mala Metallorum," *British Medical Journal* (1966): 5.
5. Paracelsus, *Four Treatises of Theophrastus von Hohenheim Called Paracelsus*, trans. George Rosen (Baltimore, MD: Johns Hopkins University Press, 1941), 57.
6. Paracelsus, *Four Treatises*, 69.
7. Emily R. Kelly, "Paracelsus the Innovator: A Challenge to Galenism from *On the Miner's Sickness and Other Miners' Diseases*," *University of Western Ontario Medical Journal* 78 (2008): 70.
8. Ball, *The Devil's Doctor*, 252–66.
9. George Rosen, introduction to *Four Treatises of Theophrastus von Hohenheim Called Paracelsus* (Baltimore, MD: Johns Hopkins University Press, 1941), 49.
10. Paracelsus, *Treatises of Theophrastus von Hohenheim Called Paracelsus*, 68.
11. Herbert Hoover and Lou Hoover, introduction to *De Re Metallica by Agricola*, trans. Herbert Hoover and Lou Hoover (New York: Dover, 1950).
12. Herbert Hoover and Lou Hoover, "Translators' Preface," in *De Re Metallica by Agricola*.
13. Agricola, *De Re Metallica*, 6.
14. Agricola, *De Re Metallica*, 214.
15. Wilmer Wright, introduction to *De Morbis Arifficum* (Chicago: University of Chicago Press, 1940).
16. Wright, introduction.
17. Ramazzini, *De Morbis Arifficum Diatriba*, trans. Wilmer Wright (Chicago: University of Chicago Press, 1940), 1.
18. Ramazzini, *De Morbis Arifficum Diatriba*, 21–45.
19. Ramazzini, *De Morbis Arifficum Diatriba*, 53.
20. Ramazzini, *De Morbis Arifficum Diatriba*, 69.
21. Andrew Meiklejohn, "History of Lung Diseases of Coal Miners in Great Britain: I. 1800–1875," *British Journal of Industrial Medicine* 8 (1951): 127–37.

22. Meiklejohn, "History of Lung Diseases of Coal Miners in Great Britain: I."
23. Andrew Meiklejohn, "Pneumoconiosis," *Postgraduate Medical Journal* (December 1949): 599–610.
24. Andrew Meiklejohn, "History of Lung Diseases of Coal Miners in Great Britain: I."
25. Andrew Meiklejohn, "History of Lung Diseases of Coal Miners in Great Britain: Part II, 1875–1920," *British Journal of Industrial Medicine* 9 (1952): 93–98.
26. Meiklejohn, "History of Lung Diseases: Part II."
27. Meiklejohn, "History of Lung Diseases: Part II."
28. J. McDonald, F. Liddell, G. Gibbs, G. Eyssen, and A. McDonald, "Dust Exposure and Mortality in Chrysotile Mining, 1910–75," *British Journal of Industrial Medicine* 37 (1980): 11–24.
29. Lundy Braun and Sophia Kisting, "Asbestos-Related Disease in South Africa: The Social Production of an Invisible Epidemic," *American Journal of Public Health* 96 (2006): 1386–96.
30. Braun and Kisting, "Asbestos-Related Disease in South Africa."
31. Andrew Meiklejohn, "History of Lung Diseases of Coal Miners in Great Britain: III. 1920–1952," *British Journal of Industrial Medicine* 9 (1952): 208–20.
32. International Agency for Research on Cancer, "Arsenic, Metals, Fibres and Dusts," *IARC Monographs on the Evaluation of Carcinogenic Risks to Humans* 100C (2012).
33. Barry Meier, "Quartz Countertops Pose a Lethal Risk to the Workers Who Fabricate Them," *New York Times*, June 2, 2016.

5. The Chemical Age

1. E. Kinne-Saffran and Rolf Kinne, "Vitalism and Synthesis of Urea from Friedrich Wöhler to Hans A. Krebs," *American Journal of Nephrology* 19 (1999): 290–94.
2. Kinne-Saffran and Kinne, "Vitalism and Synthesis of Urea."
3. Sharon Bertsch McGrayne, *Prometheans in the Lab: Chemistry and the Making of the Modern World* (New York: McGraw-Hill, 2001), 17–19.
4. McGrayne, *Prometheans in the Lab*, 17–19.
5. M. Case, Margorie Hosker, Drever McDonald, and Joan Pearson, "Tumours of the Urinary Bladder in Workmen Engaged in the Manufacture and Use of Certain Dyestuff Intermediates in the British Chemical Industry: I. The Role of Aniline, Benzidine, Alpha-Naphthylamine, and Beta-Naphthylamine," *British Journal of Industrial Medicine* 11 (1954): 75–104.
6. Christopher Sellers, "Discovering Environmental Cancer: Wilhelm Hueper, Post–World War II Epidemiology, and the Vanishing Clinician's Eye," *American Journal of Public Health* 87 (1997): 1824–35.
7. T. S. Scott, "The Incidence of Bladder Tumours in a Dyestuffs Factory," *British Journal of Industrial Medicine* 9 (1952): 127–32.
8. Case, "Tumours of the Urinary Bladder in Workmen: I," 75–104.
9. John Whysner, Lynn Verna, and Gary Williams, "Benzidine Mechanistic Data and Risk Assessment: Species- and Organ-Specific Metabolic Activation," *Pharmacology and Therapeutics* 71 (1996): 107–26.
10. H. H. Lowry, *Chemistry of Coal Utilization*, vol. 2 (New York: Wiley, 1945).

11. Howard Batchelder, *Chemicals from Coal* (Columbus, OH: Battelle Memorial Institute, Columbus Laboratories, 1970).

12. Alice Hamilton, "The Growing Menace of Benzene (Benzol) Poisoning in American Industry," *Journal of the American Medical Association* 78 (1922): 627–30.

13. Manfred Bowditch and Hervey Elkins, "Chronic Exposure to Benzene (Benzol): The Industrial Aspects," *Journal of Industrial Hygiene and Toxicology* 21 (1939): 321–77. Francis Hunter, "Chronic Exposure to Benzene (Benzol): II. The Clinical Effects," *Journal of Industrial Hygiene and Toxicology* 21 (1939): 331–54.

14. Peter Infante, Robert Rinsky, Joseph Wagoner, and Ronald Young, "Benzene and Leukaemia," *Lancet* 2 (1977): 868–69.

15. Ron Chernow, *Titan: The Life of John D. Rockefeller Sr.* (New York: Random House, 1998).

16. Batchelder, *Chemicals from Coal*.

17. International Agency for Research on Cancer, "Benzene," in *Some Anti-Thyroid and Related Substances, Nitrofurans and Industrial Chemicals*, IARC Monographs on the Evaluation of Carcinogenic Risk of Chemicals to Humans 7 (Lyon: IARC, 1974). 203–21.

18. *Toxicology: The Basic Science of Poisons*, ed. Louise J. Casarett and John Doull (New York: Macmillan, 1975).

19. Lois Travis, Chin-Yang Li, Zhi-Nan Zhang, et al., "Hematopoietic Malignancies and Related Disorders Among Benzene-Exposed Workers in China," *Leukemia and Lymphoma* 14 (1994): 91–102.

6. The Bioassay Boom

1. Leon Wiltse and T. Glenn Pait, "Herophilus of Alexandria (325–255 BC): The Father of Anatomy," *Spine Journal* 23 (1998): 1904–14.

2. Roy Porter, *The Greatest Benefit to Mankind: A Medical History of Humanity* (New York: Norton, 1997), 74–75.

3. Galen, "On the Natural Faculties," in *Great Books of the Western World*, vol. 10, *Hippocrates, Galen*, trans. Arthur John Brock (Chicago: Encyclopedia Britannica, 1952), 1:13.

4. Sanjib Ghosh, "Human Cadaveric Dissection: A Historical Account from Ancient Greece to the Modern Era," *Anatomy & Cell Biology* 48 (2015): 153–69.

5. Porter, *The Greatest Benefit to Mankind*, 211–13.

6. Jacalyn Duffin, *History of Medicine*, 2nd ed. (Toronto: University of Toronto Press, 2010), 11–32.

7. Porter, *The Greatest Benefit to Mankind*, 431–35.

8. Porter, *The Greatest Benefit to Mankind*, 436–37.

9. Carol Ballentine, "Taste of Raspberries, Taste of Death: The 1937 Elixir Sulfanilamide Incident," *FDA Consumer Magazine*, June 1981.

10. Letter by Dr. A. S. Calhoun, October 22, 1937, quoted in Ballentine, "Taste of Raspberries, Taste of Death."

11. Quoted in Ballentine, "Taste of Raspberries, Taste of Death."

12. Ballentine, "Taste of Raspberries, Taste of Death."

13. G. F. Somers, "Pharmacological Properties of Thalidomide (Alpha-Phthalimido Glutarimide), a New Sedative Hypnotic Drug," *British Journal of Pharmacology* 15 (1960): 111–16.

14. Philip Hilts, *Protecting America's Health* (New York: Knopf, 2003), 144–65.
15. James Schardein, *Chemically Induced Birth Defects*, 3rd ed. (New York: Marcel Dekker, 2000), 89–119.
16. "Thalidomide and Congenital Malformations," *Canadian Medical Association Journal* 86 (1962): 462–63.
17. "Thalidomide and Congenital Malformations," 462–63.
18. "Dr. Frances Kathleen Oldham Kelsey," *Changing the Face of Medicine*, National Institutes of Health, http://www.nlm.nih.gov/changingthefaceofmedicine/physicians/biography_182.html.
19. Hilts, *Protecting America's Health*, 144–65.
20. Pamela Fullerton and Michael Kremer, "Neuropathy After Intake of Thalidomide (Distaval)," *British Medical Journal* 2 (1961): 855–58. Hilts, *Protecting America's Health*, 144–65.
21. Schardein, *Chemically Induced Birth Defects*, 89–119.
22. Wallace F. Janssen, "The Story of the Laws Behind the Labels, Part I: The 1906 Food and Drugs Act," *FDA Consumer*, June 1981.
23. Schardein, *Chemically Induced Birth Defects*, 31–32.
24. Janssen, "The Story of the Laws Behind the Labels, Part I"; D. McFadden, "An F.D.A. Stickler Who Saved U.S. Babies from Thalidomide," *New York Times*, August 8, 2015.
25. U.S. Congress, Office of Technology Assessment, *Identifying and Regulating Carcinogens*, OTA-BP-H-42 (Washington, DC: U.S. Government Printing Office, 1987), 29–31.
26. U.S. Congress, Office of Technology Assessment, *Identifying and Regulating Carcinogens*, 147.
27. U.S. Congress, Office of Technology Assessment, *Identifying and Regulating Carcinogens*, 13.
28. U.S. Congress, Office of Technology Assessment, *Identifying and Regulating Carcinogens*, 37–39.
29. Takayuki Shibamoto and Leonard F. Bjeldanes, *Introduction to Food Toxicology* (San Diego, CA: Academic Press, 1993), 31.
30. Shibamoto and Bjeldanes, *Introduction to Food Toxicology*, 147–48.
31. Robert Cole, "Calandra Out as Bio-Test Head; Concern Under Study of F.D.A.," *New York Times*, March 26, 1977.
32. "The Scandal in Chemical Testing," *New York Times*, May 16, 1983.
33. Margot Slade and Eva Hoffman, "Ideas and Trends in Summary; Laboratory Official Accused of Fudging," *New York Times*, June 28, 1981.
34. "Three Ex-Officials of Major Laboratory Convicted of Falsifying Drug Tests," *New York Times*, October 22, 1983.
35. Richard Lyons, "Effort to Assess Pesticide Safety Is Bogged Down," *New York Times*, December 12, 1977.
36. World Health Organization, *Handbook: Good Laboratory Practice (GLP): Quality Practices for Regulated Non-Clinical Research and Development*, 2nd ed. (World Health Organization on Behalf of the Special Programme for Research and Training in Tropical Diseases, 2009), 5.
37. "U.S. Statistics," *Speaking of Research*, https://speakingofresearch.com/facts/statistics/; USDA, "Annual Report Animal Usage by Fiscal Year," fiscal year 2016,

Speaking of Research, https://speakingofresearch.files.wordpress.com/2008/03/usda-annual-report-animal-usage-in-research-2016.pdf.

38. Annamaria A. Bottini and Thomas Hartung, "Food for Thought: On the Economics of Animal Testing," *Alternatives to Animal Testing* 26 (2009): 3–19.

7. Lead: A Heavy Metal Weighing Down the Brain

1. Alice Hamilton, *Exploring the Dangerous Trades* (Beverly, MA: OEM, 1995).

2. Paul Mushak and Annemarie Crocetti, "Methods for Reducing Lead Exposure in Young Children and Other Risk Groups: An Integrated Summary of a Report to the U.S. Congress on Childhood Lead Poisoning," *Environmental Health Perspectives* 89 (1990): 125–35.

3. Evelyn Hartman, Wilford Park, and Godfrey Nelson, "The Peeling House Paint Hazard to Children," *Public Health Report* 75 (1960): 623–29.

4. Sharon Bertsch McGrayne, *Prometheans in the Lab: Chemistry and the Making of the Modern World* (New York: McGraw-Hill, 2001), 79–105.

5. McGrayne, *Prometheans in the Lab*.

6. McGrayne, *Prometheans in the Lab*.

7. Jane Lin-Fu, "Modern History of Lead Poisoning: A Century of Discovery and Rediscovery," in *Human Lead Exposure*, ed. Herbert L. Needleman (Boca Raton, FL: CRC, 1991), 31.

8. McGrayne, *Prometheans in the Lab*, 168–97.

9. McGrayne, *Prometheans in the Lab*.

10. U.S. Department of Health and Human Services, *Blood Lead Levels for Persons Ages 6 Months–74 Years: United States, 1976–80*, Department of Health and Human Services Publication no. (PHS) 84-1683.

11. J. Julian Chisolm Jr. and Donald Barltrop, "Recognition and Management of Children with Increased Lead Absorption," *Archives of Disease in Childhood* 54 (1979): 249–62.

12. Lin-Fu, "Modern History of Lead Poisoning."

13. Herbert Needleman, "The Future Challenge of Lead Toxicity," *Environmental Health Perspectives* 89 (1990): 85–89.

14. Lin-Fu, "Modern History of Lead Poisoning."

15. Ellen Silbergeld, "Implications of New Data on Lead Toxicity for Managing and Preventing Exposure," *Environmental Health Perspectives* 89 (1990): 49–54.

16. Joel Nigg, G. Mark Knottnerus, Michelle Martel, et al., "Low Blood Lead Levels Associated with Clinically Diagnosed Attention-Deficit/Hyperactivity Disorder and Mediated by Weak Cognitive Control," *Biological Psychiatry* 63 (2008): 325–31. Soon-Beom Hong, Mee-Hyang Im, Jae-Won Kim, et al., "Environmental Lead Exposure and Attention Deficit/Hyperactivity Disorder Symptom Domains in a Community Sample of South Korean School-Age Children," *Environmental Health Perspectives* 123 (2015): 271–76.

17. Brian Boutwell, Erik J. Nelson, Brett Emoc, et al., "The Intersection of Aggregate-Level Lead Exposure and Crime," *Environmental Research Letters* 148 (2016): 79–85.

18. Wayne Hall, "Did the Elimination of Lead from Petrol Reduce Crime in the USA in the 1990s? Version 2," *F1000 Research* 2 (July 16, 2013 [revised October 8, 2013]): 156.

19. Boutwell et al., "The Intersection of Aggregate-Level Lead Exposure and Crime," 79–85.

20. In 2007, the ATSDR concluded that the evidence suggested that high levels of exposure could cause an increase in miscarriages. However, the level in these workers was not as high. Agency for Toxic Substance and Disease Registry, *Toxicological Profile for Lead* (Atlanta, GA: U.S. Department of Health and Human Services, 2007).

21. Stuart Kiken, Thomas Sinks, William Stringer, Marian Coleman, Michael Crandall, and Teresa Seitz, "NIOSH Investigation of USA Today/Gannett Co. Inc.," *HETA 89-069-2036*, April 1990.

22. Chinaro Kennedy, Ellen Yard, Timothy Dignam, et al., "Blood Lead Levels Among Children Aged < 6 Years—Flint, Michigan, 2013-2016," *Morbidity and Mortality Weekly Report* 65 (2016): 650–54.

23. Michael Wines, "Flint Is in the News, but Lead Poisoning Is Even Worse in Cleveland," *New York Times*, March 3, 2016.

24. "Tests Show High Lead Levels in Water at 60 Cleveland Schools," *New York Times*, November 18, 2016.

25. Wines, "Flint Is in the News, but Lead Poisoning Is Even Worse in Cleveland."

26. Agency for Toxic Substance and Disease Registry, *Toxicological Profile for Lead*, 289–94.

27. Abby Goodnough, "Their Soil Toxic, 1,100 in Indiana Are Uprooted," *New York Times*, August 31, 2016.

28. Agency for Toxic Substance and Disease Registry, *Toxicological Profile for Lead*, 336–44.

29. Agency for Toxic Substance and Disease Registry, *Toxicological Profile for Lead*, 336–44.

30. Wines, "Flint Is in the News, but Lead Poisoning Is Even Worse in Cleveland."

8. Rachel Carson: Silent Spring Is Now Noisy Summer

1. Therese Schooley, Michael J. Weaver, Donald Mullins, and Matthew Eick, "The History of Lead Arsenate Use in Apple Production: Comparison of Its Impact in Virginia with Other States," *Journal of Pesticide Safety Education* 10 (2008): 22–53.

2. Sharon Bertsch McGrayne, *Prometheans in the Lab: Chemistry and the Making of the Modern World* (New York: McGraw-Hill, 2001), 148–67.

3. Paul Muller, "Dichloro-diphenyl-trichloroethane and Newer Insecticides—Nobel Lecture, December 11, 1948," in *Nobel Lectures: Physiology or Medicine, 1942–1962* (Amsterdam: Elsevier, 1964), 221–40.

4. McGrayne, *Prometheans in the Lab*, 148–67.

5. William Souder, *On a Farther Shore: The Life and Legacy of Rachel Carson* (New York: Crown, 2012), 332.

6. David Kinkela, *DDT and the American Century* (Chapel Hill: University of North Carolina Press, 2014), 93.

7. Souder, *On a Farther Shore*, 333.

8. Souder, *On a Farther Shore*, 245.

9. Frederick Davis, *Banned: A History of Pesticides and the Science of Toxicology* (New Haven, CT: Yale University Press, 2014), 116–21.

10. George Wallace and Richard Bernard, "Tests Show 40 Species of Birds Poisoned by DDT," *Audubon Magazine*, July/August 1963.

11. Souder, *On a Farther Shore*, 251.
12. Mark Hamilton Lytle, *The Gentle Subversive: Rachel Carson,* Silent Spring, *and the Rise of the Environmental Movement* (Oxford: Oxford University Press, 2007), 140–60.
13. Frederick Davis, "'Like a Keen North Wind': How Charles Elton Influenced *Silent Spring*," *Endeavour* 36 (2012): 143–48.
14. T. H. Jukes and C. B. Shaffer, "Antithyroid Effects of Aminotriazole," *Science* 132 (1960): 296–97.
15. Michael Tortorello, "The Great Cranberry Scare of 1959," *New Yorker*, November 24, 2015.
16. Souder, *On a Farther Shore*, 303–4.
17. John Lee, "'Silent Spring' Is Now Noisy Summer," *New York Times*, July 22, 1962.
18. Davis, *Banned*, 153–86.
19. Linda Lear, *Rachel Carson: Witness for Nature* (New York: Henry Holt, 1997), 357.
20. Rachel Carson, *Silent Spring* (Boston: Houghton Mifflin Company, 1962), 222.
21. Carson, *Silent Spring*, 222–24.
22. Davis, *Banned*, 139.
23. Souder, *On a Farther Shore*, 317.
24. Souder, *On a Farther Shore*, 352.
25. Carson, *Silent Spring*, 227–30.
26. Malcolm Hargraves, "Chemical Pesticides and Conservation Problems," lecture presented before the twenty-third annual convention of the National Wildlife Federation, February 27, 1959.
27. Lear, *Rachel Carson*, 357.
28. Carson, *Silent Spring*, 231.
29. Michelle Boland, Aparajita Chourasia, and Kay Macleod, "Mitochondrial Dysfunction in Cancer," *Frontiers in Oncology* 3 (2013): 292; Vander Heiden, L. Cantley, and C. B. Thompson, "Understanding the Warburg Effect: The Metabolic Requirements of Cell Proliferation," *Science* 324 (2009): 1029–33.
30. Lee, "'Silent Spring' Is Now Noisy Summer."
31. Robert C. Toth, "U.S. Orders Study of Two Pesticides," *New York Times*, May 5, 1963.
32. "Rachel Carson Dies of Cancer; 'Silent Spring' Author Was 56," *New York Times*, April 15, 1964.
33. Souder, *On a Farther Shore*, 335.
34. Souder, *On a Farther Shore*, 332.
35. Souder, *On a Farther Shore*, 393.
36. Souder, *On a Farther Shore*, 393–94.
37. Carson, *Silent Spring*, 225–26.
38. International Agency for Research on Cancer, "Amitrole," in *Some Thyrotropic Agents*, IARC Monographs on the Evaluation of Carcinogenic Risks to Humans 79 (Lyon: IARC, 2001).
39. Agency for Toxic Substances and Disease Registry, *DDT Toxicological Profile* (Atlanta, GA: Department of Health and Human Services, 2002).
40. Florence Breeveld, Stephen Vreden, and Martin Grobusch, "History of Malaria Research and Its Contribution to the Malaria Control Success in Suriname: A Review," *Malaria Journal* 11 (2012): 95.
41. Amir Attaran and Rajendra Maharaj, "Ethical Debate: Doctoring Malaria, Badly: The Global Campaign to Ban DDT," *British Medical Journal* 321 (2000): 1403–5.

42. Vladimir Turusov, Valery Rakitsky, and Lorenzo Tomatis, "Dichlorodiphenyltrichloroethane (DDT): Ubiquity, Persistence, and Risks," *Environmental Health Perspectives* 110 (2002): 125–28.

43. Fredric Steinberg, "Is It Time to Dismiss Calls to Ban DDT," *British Medical Journal* 322 (2001): 676–77.

44. Gretchen Vogel, "Malaria May Accelerate Aging in Birds," *Science* 347 (2015): 362.

9. The Study of Cancer

1. Percival Pott, *Chirurgical Observations Relative to the Cataract, the Polypus of the Nose, the Cancer of the Scrotum, [etc.]* (London, 1775).

2. Henry Butlin, "Three Lectures on Cancer of the Scrotum in Chimney-Sweeps and Others: Lecture 1—Secondary Cancer Without Primary Cancer," *British Medical Journal* 1 (1892): 1341–46.

3. Henry Butlin, "Three Lectures on Cancer of the Scrotum in Chimney-Sweeps and Others: Lecture 3—Tar and Paraffin Cancers," *British Medical Journal* 2 (1892): 66–71.

4. John Simmons, *Doctors and Discoveries: Lives That Created Today's Medicine* (Boston: Houghton Mifflin Harcourt, 2002), 60.

5. Howard Haggard and G. M. Smith, "Johannes Muller and the Modern Conception of Cancer," *Yale Journal of Biology and Medicine* 10 (1938): 419–36.

6. Haggard and Smith, "Johannes Muller."

7. Haggard and Smith, "Johannes Muller."

8. Henry Harris, *The Birth of the Cell* (New Haven, CT: Yale University Press, 1999).

9. Leon Bignold, Brian Coghlan, and Hubertus Jersmann, *David Paul von Hansemann: Contributions to Oncology* (Basel: Birkhauser Verlag, 2007), 41–55.

10. Bignold, Coghlan, and Jersmann, *David Paul von Hansemann*, 75–90.

11. Theodor Boveri, *Concerning the Origin of Malignant Tumours*, trans. and ed. Henry Harris (Cold Spring Harbor, NY: The Company of Biologists Limited and Cold Spring Harbor Laboratory Press, 2008).

12. Boveri, *Concerning the Origin of Malignant Tumours*.

13. Charlotte Auerbach, John Robson, and J. G. Carr, "The Chemical Production of Mutations," *Science* 105 (1947): 243–47.

14. Sverre Heim and Felix Mitelman, "A New Approach to an Old Problem, Chapter 1," in *Cancer Cytogenetics*, 3rd ed., ed. Sverre Heim and Felix Mitelman (Hoboken, NJ: Wiley-Blackwell, 2009), 1–8.

15. Joe Hin Tjio and Albert Levan, "The Chromosome Number of Man," *Hereditas* 42 (1956): 1–6.

16. Macfarlane Burnet, *The Clonal Selection Theory of Acquired Immunity* (Cambridge: Cambridge University Press, 1959).

17. James Watson, *The Double Helix* (New York: Simon and Shuster, 1968).

18. James Watson and Francis Crick, "The Structure of DNA," *Cold Spring Harbor Symposia on Quantitative Biology* 18 (1953): 123–31.

19. U.S. National Library of Medicine Profiles in Science, "Marshall W. Nirenberg," http://profiles.nlm.nih.gov/ps/retrieve/Narrative/JJ/p-nid/21.

20. U.S. National Library of Medicine Profiles in Science, "Marshall W. Nirenberg."

10. How Are Carcinogens Made?

1. Sharon Bertsch McGrayne, *Prometheans in the Lab: Chemistry and the Making of the Modern World* (New York: McGraw-Hill, 2001), 17–19.

2. Henry Butlin, "Three Lectures on Cancer of the Scrotum in Chimney-Sweeps and Others: Lecture III—Tar and Paraffin Cancer," *British Medical Journal* 2 (1892): 66–71.

3. Katsusaburo Yamagiwa and Koichi Ichikawa, "Experimental Study of the Pathogenesis of Carcinoma," *Journal of Cancer Research* 3 (1918): 1–29.

4. Rony Armon, "From Pathology to Chemistry and Back: James W. Cook and Early Chemical Carcinogenesis Research," *AMBIX* 59 (2012): 152–69.

5. Isaac Berenblum and Philippe Shubik, "The Role of Croton Oil Applications, Associated with a Single Painting of a Carcinogen, in Tumour Induction of the Mouse's Skin," *British Journal of Cancer* 1 (1947): 379–82.

6. David Clayson, *Chemical Carcinogenesis* (London: J. & A. Churchill, 1962), 410–37.

7. Clayson, *Chemical Carcinogenesis*.

8. Peter Czygan, Helmut Greim, Anthony Garro, et al., "Microsomal Metabolism of Dimethylnitrosamine and the Cytochrome P-450 Dependency of Its Activation to a Mutagen," *Cancer Research* 33 (1973): 2983–86.

9. Alvito Alvares, Gayle Schilling, Wayne Levin, and Ronald Kuntzman, "Studies on the Induction of CO-Binding Pigments in Liver Microsomes by Phenobarbital and 3-Methylcholanthrene," *Biochemical and Biophysical Research Communications* 29 (1967): 521–26.

10. Andrew Parkinson, Brian Ogilvie, David Buckley, Faraz Kazmi, Maciej Czerwinski, and Oliver Parkinson, "Biotransformation of Xenobiotics," in *Casarett and Doull's Toxicology: The Basic Science of Poisons*, 8th ed., ed. Curtis Klassen (New York: McGraw-Hill, 2013), 253.

11. Some Carcinogens Directly Affect Genes

1. Peter Brookes and Philip Lawley, "The Reaction of Mustard Gas with Nucleic Acids in Vitro and in Vivo," *Biochemical Journal* 77 (1960): 478–84.

2. Philip Lawley and Peter Brookes, "Further Studies on the Alkylation of Nucleic Acids and Their Constituent Nucleotides," *Biochemical Journal* 89 (1963): 127–38.

3. Peter Brookes and Philip Lawley, "Evidence for the Binding of Polynuclear Aromatic Hydrocarbons to the Nucleic Acids of Mouse Skin: Relation Between Carcinogenic Power of Hydrocarbons and Their Binding to Deoxyribonucleic Acid," *Nature* 202 (1964): 781–84.

4. Elizabeth Miller and James Miller, "Mechanisms of Chemical Carcinogenesis: Nature of the Proximate Carcinogens and Interactions with Macromolecules," *Pharmacological Reviews* 18 (1966): 805–38.

5. Miller and Miller, "Mechanisms of Chemical Carcinogenesis."

6. Fred Kadlubar, James Miller, and Elizabeth Miller, "Guanyl O6-Arylamination and O6-Arylation of DNA by the Carcinogen N-Hydroxy-1-Naphthylamine," *Cancer Research* 38 (1978): 3628–38. Fred Kadlubar, "A Transversion Mutation Hypothesis for Chemical Carcinogenesis by N2-Substitution of Guanine in DNA," *Chemical-Biological Interactions* 31 (1980): 255–63.

7. Bruce Ames, E. Gurney, James Miller, and Helmut Bartsch, "Carcinogens as Frameshift Mutagens: Metabolites and Derivatives of 2-Acetylaminofluorene and Other Aromatic Amine Carcinogens," *Proceedings of the National Academy of Sciences of the United States of America* 69 (1972): 3128–32.

8. Bruce Ames, P. Sims, and P. L. Grover, "Epoxides of Carcinogenic Polycyclic Hydrocarbons Are Frameshift Mutagens," *Science* 176 (1972): 47–49.

9. Bruce Ames, William Durston, Edith Yamasaki, and Frank Lee, "Carcinogens Are Mutagens: A Simple Test System Combining Liver Homogenates for Activation and Bacteria for Detection," *Proceedings of the National Academy of Sciences of the United States of America* 70 (1973): 2281–85.

10. Ames, Sims, and Grover, "Epoxides of Carcinogenic Polycyclic Hydrocarbons Are Frameshift Mutagens"; Ames, Durston, Yamasaki, and Lee, "Carcinogens Are Mutagens."

11. Erik Stokstad, "DNA's Repair Trick Win Chemistry's Top Prize," *Science* 350 (2015): 266.

12. John Whysner, M. Vijayaraj Reddy, Peter Ross, Melissa Mohan, and Elizabeth Lax, "Genotoxicity of Benzene and Its Metabolites," *Mutation Research* 566 (2004): 99–130.

13. MaryJean Pendleton, R. Hunter Lindsey Jr., Carolyn A. Felix, David Grimwade, and Neil Osheroff, "Topoisomerase II and Leukemia," *Annals of the New York Academy of Sciences* 1310 (2014): 98–110.

14. Robert Weinberg, *One Renegade Cell: How Cancer Begins* (New York: Basic Books, 1998), 25–44.

15. Weinberg, *One Renegade Cell*, 63–78. Arnold Levine, "Tumor Suppressor Genes," *Bioessays* 2 (1990): 60–66.

16. Weinberg, *One Renegade Cell*, 126–30.

17. Bert Vogelstein, Eric Fearon, Stanley Hamilton, et al., "Genetic Alterations During Colorectal-Tumor Development," *New England Journal of Medicine* 319 (1988): 525–32.

12. Cancer Caused by Irritation

1. Erwin Ackerknecht, *Rudolf Virchow and Virchow-Bibliographie 1843–1901*, ed. J. Schwalbe (New York: Arno, 1981), 98–99.

2. Ackerknecht, *Rudolf Virchow*, 98–99.

3. Fran Balkwill and Alberto Mantovani, "Inflammation and Cancer: Back to Virchow?" *Lancet* 357 (2001): 539–45.

4. Leon Bignold, Brian Coghlan, and Hubertus Jersmann, *David Paul von Hansemann: Contributions to Oncology* (Basel: Birkhauser Verlag, 2007), 60–61.

5. Internal Agency for Research on Cancer, "Shistosomes, Liver Flukes, and Helicobacter Pylori," in *Monographs on the Evaluation of Carcinogenic Risks to Humans* (Lyon: IARC, 1994), 61:45–119.

6. H. Kuper, H. O. Adami, and Dimitri Trichopoulos, "Infections as a Major Preventable Cause of Human Cancer," *Journal of International Medical Research* 248 (2000): 171–83.

7. Balkwill and Mantovani, "Inflammation and Cancer?"

8. Internal Agency for Research on Cancer, "Alcohol Drinking," in *Monographs on the Evaluation of Carcinogenic Risks to Humans* (Lyon: IARC, 1988), 44:153.

9. Stephan Padosch, Dirk Lachenmeier, and Lars Kröner, "Absinthism: A Fictitious Nineteenth-Century Syndrome with Present Impact," *Substance Abuse Treatment, Prevention, and Policy* 1 (2006): 14.

10. Internal Agency for Research on Cancer, "Consumption of Alcoholic Beverages," in *Monographs on the Evaluation of Carcinogenic Risks to Humans* (Lyon: IARC, 2012), 100E:373–499.

11. G. Pöschl and H. K. Seitz, "Alcohol and Cancer," *Alcohol and Alcoholism* 39 (2004): 155–65.

12. International Agency for Research on Cancer, "Consumption of Alcoholic Beverages," 373–499.

13. Isaac Berenblum and Philip Shubik, "The Role of Croton Oil Applications, Associated with a Single Painting of a Carcinogen, in Tumour Induction of the Mouse's Skin," *British Journal of Cancer* 1 (1947): 379–82.

14. Matthews Bradley, Victoria Taylor, Michael Armstrong, and Sheila Galloway, "Relationships Among Cytotoxicity, Lysosomal Breakdown, Chromosome Aberrations, and DNA Double-Strand Breaks," *Mutation Research* 189 (1987): 69–79.

15. Samuel Cohen and Leon Ellwein, "Cell Proliferation in Carcinogenesis," *Science* 249 (1990): 1007–11.

16. Gary Williams and John Whysner, "Epigenetic Carcinogens: Evaluation and Risk Assessment," *Experimental and Toxicological Pathology* 48 (1996): 189–95.

17. Lisa Coussens and Zena Werb, "Inflammation and Cancer," *Nature* 420 (2002): 860–67.

18. Sigmund Weitzman and Thomas Stossel, "Mutation Caused by Human Phagocytes," *Science* 212 (1981): 546–47.

19. Henry Pitot and Yvonne Dragan, "Chemical Carcinogenesis," in *Caserett and Doull's Toxicology: The Basic Science of Poisons*, 6th ed., ed. Curtis Klaassen (New York: McGraw-Hill, Medical Publishing Division, 2001), 241–320.

13. Cigarette Smoking: Black, Tarry Lungs

1. Richard Doll, "In Memoriam; Ernst Wynder 1923–1999," *American Journal of Public Health* 89 (1999): 1798–99.

2. Centers for Disease Control, "Mortality Trends for Selected Smoking-Related Cancers and Breast Cancer—United States, 1950–1990," *Mortality and Morbidity Weekly Report* 42:863–66.

3. Ernst Wynder and Evarts Graham, "Tobacco Smoking as a Possible Etiologic Factor in Bronchiogenic Carcinoma; A Study of 684 Proved Cases," *Journal of the American Medical Association* 143 (1950): 329–36.

4. Siddhartha Mukherjee, *The Emperor of All Maladies: A Biography of Cancer* (New York: Scribner, 2010), 244.

5. Wynder and Graham, "Tobacco Smoking as a Possible Etiologic Factor in Bronchiogenic Carcinoma."

6. Wynder and Graham, "Tobacco Smoking as a Possible Etiologic Factor in Bronchiogenic Carcinoma."

7. Ernst Wynder, "The Past, Present, and Future of the Prevention of Lung Cancer," *Cancer Epidemiology, Biomarkers & Prevention* 7 (1998): 735–48.

8. Michael Thun, "When Truth Is Unwelcome: First Reports of Smoking and Lung Cancer," *Bulletin of the World Health Organization* 83 (2005): 144–45.

9. Richard Doll and Austin Bradford Hill, "Smoking and Carcinoma of the Lung; Preliminary Report," *British Medical Journal* 2 (1950): 739–48.

10. Richard Doll and Austin Bradford Hill, "The Mortality of Doctors in Relation to Their Smoking Habits; A Preliminary Report," *British Medical Journal* 1 (1954): 1451–55.

11. Richard Doll and Austin Bradford Hill, "Lung Cancer and Other Causes of Death in Relation to Smoking; A Second Report on the Mortality of British Doctors," *British Medical Journal* 2 (1956): 1071–81.

12. Jerome Cornfield, William Haenszel, E. Cuyler Hammond, Abraham M. Lilienfeld, Michael B. Shimkin, and Ernst L. Wynder, "Smoking and Lung Cancer: Recent Evidence and a Discussion of Some Questions," *International Journal of Epidemiology* 38 (2009): 1175–91.

13. Nicole Fields and Simon Chapman, "Chasing Ernst L. Wynder: 40 Years of Philip Morris' Efforts to Influence a Leading Scientist," *Journal of Epidemiology and Community Health* 57 (2003): 571–78.

14. Ernest Wynder, Evarts Graham, and Adele Croninger, "Experimental Production of Carcinoma with Cigarette Tar," *Cancer Research* 13 (1953): 855–64.

15. Robert Weinberg, *Racing to the Beginning of the Road: The Search for the Origin of Cancer* (New York: Harmony, 1996).

16. U.S. Public Health Service, *Smoking and Health: Report of the Advisory Committee to the Surgeon General of the Public Health Service* (Princeton, NJ: D. Van Nostrand Company, 1964).

17. U.S. Public Health Service, *Smoking and Health*.

18. Steve Stellman, "Ernst Wynder: A Remembrance," *Preventive Medicine* 43 (2006): 239–45.

19. Stellman, "Ernst Wynder."

20. Annamma Augustine, Randall Harris, and Ernst Wynder, "Compensation as a Risk Factor for Lung Cancer in Smokers Who Switch from Nonfilter to Filter Cigarettes," *American Journal of Public Health* 79 (1989): 188–91.

14. What Causes Cancer?

1. Ernst Wynder and Gio Gori, "Contribution of the Environment to Cancer Incidence: An Epidemiologic Exercise," *Journal of the National Cancer Institute* 58 (1977): 825–32.

2. Richard Doll and Richard Peto, "The Causes of Cancer: Quantitative Estimates of Avoidable Risks of Cancer in the United States Today," *Journal of the National Cancer Institute* 66 (1981): 1193–308.

3. Henry Pitot and Yvonne Dragan, "Chemical Carcinogenesis," in *Casarett and Doull's Toxicology: The Basic Science of Poisons*, 6th ed., ed. Curtis D. Klaassen (New York: McGraw-Hill, 2001).

4. William Blot and Robert Tarone, "Doll and Peto's Quantitative Estimates of Cancer Risks: Holding Generally True for 35 Years," *Journal of the National Cancer Institute* 107 (2015).

5. Wynder and Gori, "Contribution of the Environment to Cancer Incidence"; Doll and Peto, "The Causes of Cancer."

6. Wynder and Gori, "Contribution of the Environment to Cancer Incidence"; Doll and Peto, "The Causes of Cancer."

7. Ernst Wynder, John Weisburger, and Stephen Ng, "Nutrition: The Need to Define 'Optimal' Intake as a Basis for Public Policy Decisions," *American Journal of Public Health* 82 (1992): 346–50.

8. Pitot and Dragan, "Chemical Carcinogenesis."

9. Béatrice Lauby-Secretan, Chiara Scoccianti, Dana Loomis, Yann Grosse, Franca Bianchini, and Kurt Straif, "Body Fatness and Cancer—Viewpoint of the IARC Working Group," *New England Journal of Medicine* 25 (2016): 794–98.

10. Tim Byers and Rebecca Sedjo, "Body Fatness as a Cause of Cancer: Epidemiologic Clues to Biologic Mechanisms," *Endocrine Related Cancer* 22 (2015): R125–34.

11. Lauby-Secretan et al., "Body Fatness and Cancer."

12. Vincent Cogliano, Robert Baan, Kurt Straif, et al., "Preventable Exposures Associated with Human Cancers," *Journal of the National Cancer Institute* 103 (2011): 1827–39.

13. Claire Vajdic, Stephen McDonald, Margaret McCredie, et al., "Cancer Incidence Before and After Kidney Transplantation," *Journal of the American Medical Association* 296 (2006): 2823–31.

14. Bernardo Ramazzini, *De Morbis Arifficum* (Chicago: University of Chicago Press, 1940), 191.

15. Internal Agency for Research on Cancer, "Post-Menopausal Oestrogen Therapy," *Hormonal Contraception and Postmenopausal Hormonal Therapy*, Monographs on the Evaluation of Carcinogenic Risks to Humans 72 (Lyon: IARC, 1999), 407.

16. Brian MacMahon, Phillip Cole, T. Lin, et al., "Age at First Birth and Breast Cancer Risk," *Bulletin of the World Health Organization* 43 (1970): 209–21.

17. D. N. Rao, B. Ganesh, and P. B. Desai, "Role of Reproductive Factors in Breast Cancer in a Low-Risk Area: A Case-Control Study," *British Journal of Cancer* 70 (1994): 129–32.

18. Mariana Chavez-MacGregor, Sjoerd Elias, Charlotte Onland-Moret, et al., "Postmenopausal Breast Cancer Risk and Cumulative Number of Menstrual Cycles," *Cancer Epidemiology Biomarkers and Prevention* 4 (2005): 799–804.

19. Margot Cleary and Michael Grossmann, "Minireview: Obesity and Breast Cancer: The Estrogen Connection," *Endocrinology* 150 (2009): 2537–42.

20. Yann Grosse, Robert Baan, Kurt Straif, et al., "A Review of Human Carcinogens—Part A: Pharmaceuticals," *Lancet Oncology* 10 (2009): 13–14.

21. Gina Kolata, "A Tradition of Caution: Confronting New Ideas, Doctors Often Hold on to the Old," *New York Times*, May 10, 1992.

22. Doll and Peto, "The Causes of Cancer."

23. F. Carneiro, *World Cancer Report: International Agency for Research on Cancer*, ed. Bernard Stewart and Christopher Wild (Lyon: IARC, 2014), 1101.

24. http://monographs.iarc.fr/ENG/Classification/Table4.pdf.

25. http://monographs.iarc.fr/ENG/Classification/Table4.pdf.

26. National Cancer Institute, *SEER Cancer Statistics Review, 1975–2012*, table 2.7, "All Cancer Sites (Invasive)," http://seer.cancer.gov/csr/1975_2012/browse_csr.php?sectionSEL=2&pageSEL=sect_02_table.07.html.

27. Bert Vogelstein, Eric Fearon, Stanley Hamilton, et al., "Genetic Alterations During Colorectal-Tumor Development," *New England Journal of Medicine* 319 (1988): 525–32.
28. Christian Tomasetti and Bert Vogelstein, "Cancer Etiology. Variation in Cancer Risk Among Tissues Can Be Explained by the Number of Stem Cell Divisions," *Science* 347 (2015): 78–81.
29. Doll and Peto, "The Causes of Cancer."
30. National Cancer Institute, "Genetic Testing for Inherited Cancer Suseptibility Syndromes," http://www.cancer.gov/about-cancer/causes-prevention/genetics/genetic-testing -fact-sheet.
31. Ernst Wynder, "American Health Foundation, 25th Anniversary Symposium," *Preventive Medicine* 25 (1996): 1–67.

15. Protecting Workers from Chemical Diseases

1. Gold Rush Trading Post, "Brief History of Drilling and Blasting," February 15, 2014, http://www.goldrushtradingpost.com/prospecting_blog/view/32702/brief_history _of_drilling_and_blasting.
2. Tim Carter, "British Occupational Hygiene Practice 1720–1920," *Annals of Occupational Hygiene* 48 (2004): 299–307.
3. "Alice Hamilton," *Mortality and Morbidity Weekly Report* 48 (1999): 462.
4. Alice Hamilton, *Exploring the Dangerous Trades* (Beverly, MA: OEM, 1995), 114–18.
5. Hamilton, *Exploring the Dangerous Trades*, 114–18.
6. Hamilton, *Exploring the Dangerous Trades*, 118–24.
7. Hamilton, *Exploring the Dangerous Trades*, 118–24.
8. Hamilton, *Exploring the Dangerous Trades*, 183–99.
9. Hamilton, *Exploring the Dangerous Trades*, 279–82.
10. Hamilton, *Exploring the Dangerous Trades*, 255–61.
11. J. C. Bridge, "The Influence of Industry on Public Health," *Proceedings of the Royal Society of Medicine* 26 (1933): 943–51.
12. ACGIH, "History," http://www.acgih.org/about-us/history.
13. U.S. Department of Labor, "Timeline of OSHA's 40 Year History," https://www.osha .gov/osha40/timeline.html.
14. David Michaels and Celeste Monforton, "Beryllium's Public Relations Problem: Protecting Workers When There Is No Safe Exposure Level," *Public Health Reports* 123 (2008): 79–88.
15. Richard Sawyer and Lisa Maier, "Chronic Beryllium Disease: An Updated Model Interaction Between Innate and Acquired Immunity," *Biometals* 24 (2011): 1–17.
16. McAllister Hull and Amy Bianco, *Rider of the Pale Horse* (Albuquerque: University of New Mexico Press, 2005).
17. Marc Stockbauer, "The Designs of Fat Man and Little Boy," EDGE: Ethics of Development in a Global Environment seminar series, Stanford University, 1999, https://web .stanford.edu/class/e297c/war_peace/atomic/hfatman.html.
18. W. Jones Williams, "A Histological Study of the Lungs in 52 Cases of Chronic Beryllium Disease," *British Journal of Industrial Medicine* 15 (1958): 84–91.
19. Dannie Middleton, "Chronic Beryllium Disease: Uncommon Disease, Less Common Diagnosis," *Environmental Health Perspectives* 106 (1998): 765–67.

20. Kenneth Rosenman, Vicki Hertzberg, Carol Rice, et al., "Chronic Beryllium Disease and Sensitization at a Beryllium Processing Facility," *Environmental Health Perspective* 113 (2005): 1366–72.

21. Dan Middleton and Peter Kowalski, "Advances in Identifying Beryllium Sensitization and Disease," *International Journal of Environmental Research and Public Health* 7 (2010): 115–24.

22. Middleton and Kowalski, "Advances in Identifying Beryllium Sensitization and Disease."

23. U.S. Department of Labor, "Final Rule to Protect Workers from Beryllium Exposure," December 10, 2018, https://www.osha.gov/berylliumrule/index.html.

16. The Importance of Having a Good Name

1. International Agency for Research on Cancer, "Agents Classified by the *IARC Monographs*, Volumes 1–23," November 2, 2018, http://monographs.iarc.fr/ENG/Classification/ClassificationsAlphaOrder.pdf.

2. IARC, "Agents Classified by the *IARC Monographs*."

3. Girard Hottendorf and Irwin Pachter, "Review and Evaluation of the NCI/NTP Carcinogenesis Bioassays," *Toxicologic Pathology* 13 (1985): 141–46.

4. Peter Shields, John Whysner, and Kenneth Chase, "Polychlorinated Biphenyls and Other Polyhalogenated Aromatic Hydrocarbons," in *Hazardous Materials Toxicology*, ed. J. Sullivan (Baltimore, MD: Williams & Wilkins, 1992).

5. William Blair, "Senate Unit Told of Fish Tainting, Chemical Is the Same Found in Chickens Near Factory," *New York Times*, August 5, 1971.

6. Kevin Sack, "PCB Pollution Suits Have Day in Court in Alabama," *New York Times*, January 27, 2002.

7. Blair, "Senate Unit Told of Fish Tainting."

8. Renata Kimbrough, Robert Squire, R. E. Linder, John Strandberg, R. J. Montalli, and Virlyn Burse, "Induction of Liver Tumor in Sherman Strain Female Rats by Polychlorinated Biphenyl Aroclor 1260," *Journal of the National Cancer Institute* 55 (1975): 1453–59.

9. Jacques Steinberg, "The 13-Year Cleaning Job; After $53 Million, a $17 Million State Building Finally Is Declared Safe from Toxins," *New York Times*, October 11, 1994.

10. Richard Severo, "State Says Some Striped Bass and Salmon Pose a Toxic Peril," *New York Times*, August 8, 1975.

11. American Lung Association, "Tobacco Initiatives," https://www.lung.org/our-initiatives/tobacco/.

17. Can We Accurately Regulate Chemicals?

1. International Agency for Research on Cancer, "Benzene," in *Some Anti-Thyroid and Related Substances, Nitrofurans and Industrial Chemicals*, IARC Monographs on the Evaluation of Carcinogenic Risk of Chemicals to Humans 7 (Lyon: IARC, 1974), 203–21.

2. Enrico Vigliani, "Leukemia Associated with Benzene Exposure," *Annals of the New York Academy of Sciences* 271 (1976): 143–51. Enrico Vigliani and Giulio Saita, "Benzene and Leukemia," *New England Journal of Medicine* 271 (1964): 872–76.

3. Austin Bradford Hill, "The Environment and Disease: Association or Causation?" *Proceedings of the Royal Society of Medicine* 58 (1965): 295–300.
4. Hill, "The Environment and Disease."
5. Hill, "The Environment and Disease."
6. Hill, "The Environment and Disease."
7. Hill, "The Environment and Disease."
8. Carolyn Raffensperger and Joel Tickner, introduction to *Protecting Public Health and the Environment: Implementing the Precautionary Principle*, ed. Carolyn Raffensperger and Joel Tickner (Washington, DC: Island, 1999), 1–11.
9. Raffensperger and Tickner, introduction.
10. International Agency for Research on Cancer, "Preamble," in *Occupational Exposures to Mists and Vapours from Strong Inorganic Acids; and Other Industrial Chemicals*, IARC Monographs on the Evaluation of Carcinogenic Risk of Chemicals to Humans 54 (Lyon: IARC, 1992).
11. International Agency for Research on Cancer, "IQ," "MOCA," "Methylmercury Compounds," "Ethylene Oxide," "Styrene," "Acylamide," "Trichloropropane," "Vinyl Acetate," "Vinyl Fluoride," "Dioxin," "Aziridine," "Diethyl Sulfate," 'Dimethy-carbamoyl Chloride," "Diemthylhydrazine," "Dimethylsulfate," "Epichlorlydrin," "Epoxybutane," "Ethylene Dibromide," "Methyl Methanesulfonate," "Tris Dibro-mopropyl Phosphate," and "Vinylbromide," IARC Monographs on the Evaluation of Carcinogenic Risk of Chemicals to Humans 54–72 (Lyon: IARC, 1992–1998).
12. U.S. Environmental Protection Agency, *Alpha 2u-Globulin: Association with Chemically Induced Renal Toxicity and Neoplasia in the Male Rat* (Washington, DC: Risk Assessment Forum, 1991), EPA/625/3-91/019F.
13. Charles Capen, Erik Dybing, Jerry Rice, and Julian Wilbourn, "Consensus Report," in *Species Differences in Thyroid, Kidney, and Urinary Bladder Carcinogenesis*, IARC Scientific Publications 147 (Lyon: IARC, 1999).
14. Dan Ferber, "Lashed by Critics, WHO's Cancer Agency Begins a New Regime," *Science* 301 (2003): 36–37.
15. Vincent Cogliano, Robert Baan, Kurt Straif, Yann Grosse, Beatrice Secretan, and Fatiha El Ghissassi, "Use of Mechanistic Data in IARC Evaluations," *Environmental and Molecular Mutagenesis* 49 (2008): 100–9.

18. The Dose Makes the Poison

1. International Agency for Research on Cancer, "Paracetamol," in *Some Chemicals That Cause Tumours of the Kidney or Urinary Bladder in Rodents and Some Other Substances*, IARC Monographs on the Evaluation of Carcinogenic Risk of Chemicals to Humans 73 (Lyon: IARC, 1999), 401–50.
2. International Agency for Research on Cancer, "Paracetamol."
3. Michael Fleming, S. John Mihic, and R. Adron Harris, "Ethanol," in *Goodman & Gillman's Pharmacological Basis of Therapeutics*, 11th ed., ed. Laurence Brunton, John Lazo, and Keith Parker (New York: McGraw-Hill, 2006), 693.
4. Sook Young Lee, "Can Liver Toxicity Occur at Repeated Borderline Supratherapeutic Doses of Paracetamol?" *Hong Kong Medical Journal* 10 (2004): 220, 221–22.
5. Anne Burke, Emer Smyth, and Garret FitzGerald, "Analgesic-Antipyretic and Anti-Inflammatory Agents; Pharmacotherapy of Gout," in *Goodman & Gillman's*

Pharmacological Basis of Therapeutics, 11th ed., ed. Laurence Brunton, John Lazo, and Keith Parker (New York: McGraw-Hill, 2006), 593.

6. Fleming, Mihic, and Harris, "Ethanol."

7. Barry Rumack and Frederick Lovejoy Jr., "Clinical Toxicology," in *Casarett and Doull's Toxicology: The Basic Science of Poisons*, 3rd ed., ed. Curtis Klaassen, Mary Amdur, and John Doull (New York: Macmillan, 1986), 879–901.

8. Edward Calabrese, Molly McCarthy, and Elaina Kenyon, "The Occurrence of Chemically Induced Hormesis," *Health Physics* 52 (1987): 531–41; Edward Calabrese and Linda Baldwin, "Hormesis as a Biological Hypothesis," *Environmental Health Perspectives* 106 (1998): 357–62; Edward Calabrese and Linda Baldwin, "Can the Concept of Hormesis Be Generalized to Carcinogenesis?" *Regulatory Toxicolology and Pharmacology* 28 (1998): 230–41.

9. Shoji Fukushima, Anna Kinoshita, Rawiwan Puatanachokchai, Masahiko Kushida, Hideki Wanibuchi, and Keiichirou Morimura, "Hormesis and Dose-Response-Mediated Mechanisms in Carcinogenesis: Evidence for a Threshold in Carcinogenicity of Non-Genotoxic Carcinogens," *Carcinogenesis* 26 (2005): 1835–45.

10. Henry Pitot, Thomas Goldsworthy, Susan Moran, et al., "A Method to Quantitate the Relative Initiating and Promoting Potencies of Hepatocarcinogenic Agents in Their Dose-Response Relationships to Altered Hepatic Foci," *Carcinogenesis* 8 (1987): 1491–99.

11. Bureau of Labor Statistics, "Table 6: Incidence Rates and Numbers of Nonfatal Occupational Illnesses by Major Industry Sector, Category of Illness, and Ownership, 2014," updated August 27, 2016, https://www.bls.gov/news.release/osh.t06.htm.

12. S. B. Avery, D. M. Stetson, P. M. Pan, and K. P. Mathews, "Immunological Investigation of Individuals with Toluene Diisocyanate Asthma," *Clinical and Experimental Immunology* 4 (1969): 585–96; W. G. Adams, "Long-Term Effects on the Health of Men Engaged in the Manufacture of Toluene Di-Isocyanate," *British Journal of Industrial Medicine* 32 (1975): 72–78.

13. Manfred Bowditch and Hervey Elkins, "Chronic Exposure to Benzene (Benzol). I: The Industrial Aspects," *Journal of Industrial Hygiene and Toxicology* 21 (1939): 321–77; Francis Hunter, "Chronic Exposure to Benzene (Benzol). II: The Clinical Effects," *Journal of Industrial Hygiene and Toxicology* 21 (1939): 331–54.

14. Robert Rinsky, Ronald Young, and Alexander Smith, "Leukemia in Benzene Workers," *American Journal of Industrial Medicine* 2 (1981): 217–45.

15. Robert Rinsky, Alexander Smith, Richard Hornung, et al., "Benzene and Leukemia: An Epidemiologic Risk Assessment," *New England Journal Medicine* 316 (1987): 1044–50.

16. Agency for Toxic Substances and Disease Registry, *Toxicological Profile for Asbestos* (Atlanta: U.S. Department of Health and Human Services, 2001), 146.

17. D. E. Fletcher, "A Mortality Study of Shipyard Workers with Pleural Plaques," *British Journal of Industrial Medicine* 29 (1972): 142–45.

18. Richard Doll, "Mortality from Lung Cancer in Asbestos Workers," *British Journal of Industrial Medicine* 12 (1955): 81–86.

19. Cuyler Hammond, Irving Selikoff, and Herbert Seidman, "Asbestos Exposure, Cigarette Smoking, and Death Rates," *New York Academy of Sciences* 330 (1979): 473–90.

20. Harri Vainio and Paolo Boffetta, "Mechanisms of the Combined Effect of Asbestos and Smoking in the Etiology of Lung Cancer," *Scandinavian Journal of Work, Environment, and Health* 20 (1994): 235–42.

21. Sarah Huang, Marie-Claude Jaurand, David Kamp, John Whysner, and Tom Hei, "Role of Mutagenicity in Asbestos Fiber-Induced Carcinogenicity and Other Diseases," *Journal of Toxicology and Environmental Health Part B* 14 (2011): 179–245.

22. John Hedley-Whyte and Debra Milamed, "Asbestos and Ship-Building: Fatal Consequences," *Ulster Medical Journal* 77 (2008): 191–200.

23. G. Berry, M. L. Newhouse, and P. Antonis, "Combined Effect of Asbestos and Smoking on Mortality from Lung Cancer and Mesothelioma in Factory Workers," *British Journal of Industrial Medicine* 42 (1985): 12–18.

24. J. C. McDonald and A. D. McDonald, "The Epidemiology of Mesothelioma in Historical Context," *European Respiratory Journal* 9 (1996): 1932–42.

19. Are We Ready to Clean Up the Mess?

1. Kenneth Chase, Otto Wong, David Thomas, B. W. Berney, and Robert Simon, "Clinical and Metabolic Abnormalities Associated with Occupational Exposure to Polychlorinated Biphenyls (PCBs)," *Journal of Occupational Medicine* 24 (1982): 109–14.

2. U.S. Environmental Protection Agency, *Health Effects Assessment for Polychlorinated Biphenyls* (Washington, DC: U.S Environmental Protection Agency, 1984), NTIS PB 81-117798.

3. John Whysner and Gary Williams, "International Cancer Risk Assessment: The Impact of Biologic Mechanisms," *Regulatory Toxicology and Pharmacology* 15 (1992): 41–50.

4. Brian Mayes, E. McConnell, B. Neal, et al., "Comparative Carcinogenicity in Sprague-Dawley Rats of the Polychlorinated Biphenyl Mixtures Aroclors 1016, 1242, 1254, and 1260," *Toxicological Sciences* 41 (1998): 62–76.

5. John Whysner and C.-X. Wang, "Hepatocellular Iron Accumulation and Increased Proliferation in Polychlorinated Biphenyl-Exposed Sprague-Dawley Rats and the Development of Hepatocarcinogenesis," *Toxicological Sciences* 62 (2001): 36–45.

6. Food Safety Council, "Quantitative Risk Assessment," *Food and Cosmetic Toxicology* 18 (1980): 711–84; Federal Department of Agriculture, *Federal Register* 50 (1985): 45532; National Research Council, "Risk Assessment in the Federal Government: Managing the Process" (1983): 57; R. A. Tucker, "History of the Food and Drug Administration," interview with Donald Kennedy, June 17, 1996.

7. Linda Bren, "Animal Health and Consumer Protection," *FDA Consumer Magazine*, January/February 2006.

8. Nathan Mantel and W. Ray Bryan, "'Safety' Testing of Carcinogenic Agents," *Journal of the National Cancer Institute* 27 (1961): 455–70.

9. U.S. Congress, Office of Technology Assessment Task Force, *Identifying and Regulating Carcinogens*, OTA-BP-H-42 (Washington, DC: U.S. Government Printing Office, 1987), 30.

10. Katharyn Kelly, "The Myth of 10^{-6} as a Definition of Acceptable Risk (or, 'In Hot Pursuit of Superfund's Holy Grail')," presented at the 84th Annual Meeting of the Air and Waste Management Association, Vancouver, Canada, June 1991.

11. Office of Science and Technology, Office of Water, "Methodology for Deriving Ambient Water Quality Criteria for the Protection of Human Health" (Washington, DC: United States Environmental Protection Agency, 2000), 2–6, EPA-822-B-00-004.

12. John Whysner, Marvin Kushner, Vincent Covello, et al., "Asbestos in the Air of Public Buildings: A Public Health Risk?" *Preventive Medicine* 23 (1994): 119–25.

13. Health Effects Institute, *Asbestos in Public and Commercial Buildings: A Literature Review and Synthesis of Current Knowledge* (Cambridge, MA: Health Effects Institute, 1991), chap. 6.

14. Sam Dillon, "Asbestos in the Schools; Disorder on Day 1 in New York Schools," *New York Times*, September 21, 1993.

15. Sam Dillon, "Last School in Asbestos Cleanup Is to Reopen Today in Brooklyn," *New York Times*, November 18, 1993.

20. Legal Battles

1. Howard Zonana, "Daubert V. Merrell Dow Pharmaceuticals: A New Standard for Scientific Evidence in the Courts?" *Bulletin of the American Academy of Psychiatry and the Law* 22 (1994): 309–25.

2. Jose Ramon Bertomeu-Sanchez, "Popularizing Controversial Science: A Popular Treatise on Poisons by Mateu Orfila (1818)," *Medical History* 53 (2009): 351–78.

3. Bertomeu-Sanchez, "Popularizing Controversial Science."

4. Gale Cengage, "Gross, Hans," in *World of Forensic Science*, ed. K. Lee Lerner and Brenda Wilmoth Lerner (Farmington Hills, MI: Thomson Gale, 2006).

5. Marcia Angell, *Science on Trial* (New York: Norton, 1996), 125–27.

6. Zonana, "Daubert V. Merrell Dow Pharmaceuticals."

7. Zonana, "Daubert V. Merrell Dow Pharmaceuticals."

8. Janet Raloff, "Benched Science: Increasingly, Judges Decide What Science—If Any—a Jury Hears," *Science News* 168 (2005): 232–34.

9. Laurence Riff, "Daubert at 10: A View from Counsel's Table," *Inside EPA's Risk Policy Report*, December 9, 2003.

10. Raloff, "Benched Science."

11. Suzanne Orfino, "Daubert v. Merrell Dow Pharmaceuticals, Inc.: The Battle Over Admissibility Standards for Scientific Evidence in Court," *Journal of Undergraduate Science* 3 (1996): 109–11.

12. Angell, *Science on Trial*, 35–49.

13. Angell, *Science on Trial*, 52–56.

14. Angell, *Science on Trial*, 60–69.

15. Y. Okano, M. Nishikai, and A. Sato, "Scleroderma, Primary Biliary Cirrhosis, and Sjögren's Syndrome After Cosmetic Breast Augmentation with Silicone Injection: A Case Report of Possible Human Adjuvant Disease," *Annals of the Rheumatic Diseases* 43 (1984): 520–22.

16. Independent Advisory Committee on Silicone-Gel-filled Implants, "Summary of the Report on Silicone-Gel-Filled Breast Implants," *Canadian Medical Association Journal* 147 (1992): 1141–46.

17. Sherine Gabriel, W. Michael O'Fallon, Leonard Kurland, C. Mary Beard, John Woods, and Joseph Melton III, "Risk of Connective-Tissue Diseases and Other Disorders After Breast Implantation," *New England Journal of Medicine* 330 (1994): 1697–702.

18. Sherine Gabriel, W. Michael O'Fallon, C. Mary Beard, Leonard Kurland, John Woods, and Joseph Melton III, "Trends in the Utilization of Silicone Breast Implants,

1964–1991, and Methodology for a Population-Based Study of Outcomes," *Journal of Clinical Epidemiology* 48, no. 4 (1995): 527–37.

19. Laural Hooper, Joe Cecil, and Thomas Willging, "Assessing Causation in Breast Implant Litigation: The Role of Science Panels," *Law and Contemporary Problems* 64 (2001): 140–87.

20. David Kaye and Joseph Sanders, "Expert Advice on Silicone Implants: Hall v. Baxter Healthcare Corp.," *Jurimetrics Journal* 37 (1997): 113–28.

21. Jane Brody, "Shadow of Doubt Wipes Out Bendectin," *New York Times*, June 19, 1983.

22. Cynthia Crowson, Eric Matteson, Elena Myasoedova, et al., "The Lifetime Risk of Adult-Onset Rheumatoid Arthritis and Other Inflammatory Autoimmune Rheumatic Diseases," *Arthritis & Rheumatism* 63 (2011): 633–39.

23. *Brown v. SEPTA*, U.S. District Court for the Eastern District of Pennsylvania, Civil Action no. 86-2229.

24. Agency for Toxic Substances and Disease Registry, *Exposure Study of Persons Possibly Exposed to Polychlorinated Biphenyls in Paoli, Pennsylvania* (Atlanta, GA: Centers for Disease Control, 1987).

21. The Toxicology of War

1. Joel Vilensky, *Dew of Death: The Story of Lewisite, America's World War I Weapon of Mass Destruction* (Bloomington: Indiana University Press, 2005), 13–18.

2. Institute of Medicine, *Veterans at Risk: The Health Effects of Mustard Gas and Lewisite*, ed. Constance M. Pechura and David P. Rall (Washington, DC: National Academy Press, 1993), 9.

3. Frederick Sidell and Charles Hurst, "Long-Term Health Effects of Nerve Agents and Mustard," in *Textbook of Military Medicine Part: Medical Aspects of Chemical and Biological Warfare*, ed. Frederick Sidell, Ernest Takafuji, and David Franz (Washington, DC: Office of the Surgeon General, Department of the Army, United States of America, 1997).

4. Vilensky, *Dew of Death*, 13–18.

5. Lina Grip and John Hart, *The Use of Chemical Weapons in the 1935–36 Italo-Ethiopian War* (Stockholm: Stockholm International Peace Research Institute, 2009).

6. Institute of Medicine, *Veterans at Risk*, 33, 40–41.

7. Bruno Papirmeister, Alan Feister, Sabina Robinson, and Robert Ford, *Medical Defense Against Mustard Gas: Toxic Mechanisms and Pharmacological Implications* (Boca Raton, FL: CRC, 1991), 13–32.

8. Institute of Medicine, *Veterans at Risk*, 36–40.

9. Robert Joy, "Historical Aspects of Medical Defense Against Chemical Warfare," in *Textbook of Military Medicine: Medical Aspects of Chemical and Biological Warfare*, ed. Frederick Sidell, Ernest Takafuji, and David Franz (Washington, DC: Office of the Surgeon General, Department of the Army, United States of America, 1997).

10. Frederick Sidell, "A History of Human Studies with Nerve Agents by the UK and USA," in *Chemical Warfare Agents: Toxicology and Treatment*, ed. Timothy Marrs, Robert Maynard, and Frederick Sidell (London: John Wiley & Sons, 2007), 223–40.

11. Nancy Munro, Kathleen Ambrose, and Annetta Watson, "Toxicity of the Organophosphate Chemical Warfare Agents GA, GB, and VX: Implications for Public Protection," *Environmental Health Perspectives* 102 (1994): 18–38.

12. Frederick Sidell, "Nerve Agents," in *Textbook of Military Medicine: Medical Aspects of Chemical and Biological Warfare*, ed. Frederick Sidell, Ernest Takafuji, and David Franz (Washington, DC: Office of the Surgeon General, Department of the Army, United States of America, 1997).

13. Munro et al., "Toxicity of the Organophosphate Chemical Warfare Agents GA, GB, and VX."

14. Sidell, "Nerve Agents."

15. Sidell, "Nerve Agents."

16. Ulf Schmidt, *Secret Science: A Century of Poison Warfare and Human Experiments* (Oxford: Oxford University Press, 2015).

17. "Nixon Reported Set to Ban Gases," *New York Times*, November 25, 1969; John Finney, "Senate Committee Votes 1925 Chemical War Ban," *New York Times*, December 13, 1974.

18. A. P. Watson and G. D. Griffin, "Toxicity of Vesicant Agents Scheduled for Destruction by the Chemical Stockpile Disposal Program," *Environmental Health Perspectives* 98 (1992): 259–80.

19. Tim Bullman and Han Kang, "A Fifty Year Mortality Follow-Up Study of Veterans Exposed to Low Level Chemical Warfare Agent, Mustard Gas," *Annals of Epidemiology* 10 (2000): 333–38.

20. Richard Stone, "Chemical Martyrs," *Science* 359 (2018): 21–25. Steven Erlander, "A Weapon Seen as Too Horrible, Even in War," *New York Times*, September 6, 2013.

21. National Academy of Sciences, *Gulf War and Health*, vol. 1: *Depleted Uranium, Sarin, Pyridostigmine Bromide, Vaccines*, ed. Carolyn Fulco, Catharyn Liverman, and Harold Sox (Washington, DC: National Academy Press, 2000), 191.

22. Tetsu Okumura, Nobukatsu Takasu, Shinichi Ishimatsu, et al. "Report on 640 Victims of the Tokyo Subway Sarin Attack," *Annals of Emergency Medicine* 28 (1996): 129–35.

23. Edward Wong, "U.S. Says Assad May Be Using Chemical Weapons in Syria Again," *New York Times*, May 21, 2019.

22. Opiates and Politics

1. Rose Rudd, Noah Aleshire, Jon Zibbell, and R. Matthew Gladden, "Increases in Drug and Opioid Overdose Deaths—United States, 2000–2014," *Morbidity and Mortality Weekly Report* 64 (2016): 1378–82.

2. Robert Service, "New Pain Drugs May Lower Overdose and Addiction Risk," *Science* 361 (2018): 831.

3. Mike Stobbe, "Today's Opioid Crisis Shares Chilling Similarities with Past Drug Epidemics," *Chicago Tribune*, October 28, 2017.

4. Michael Brownstein, "A Brief History of Opiates, Opioid Peptides, and Opioid Receptors," *Proceedings of the National Academy of Sciences, USA* 90 (1993): 5391–93.

5. Solomon Snyder, *Brainstorming: The Science and Politics of Opiate Research* (Cambridge, MA: Harvard University Press, 1989), 32.

6. Snyder, *Brainstorming*, 38.

7. David Courtwright, *Dark Paradise: Opiate Addiction in America Before 1940* (Cambridge, MA: Harvard University Press, 1982), 46–59.

8. Arnold Trebach, *The Heroin Solution*, 2nd ed. (Bloomington, IN: Unlimited Publishing, 2006), 37–42.

9. Courtwright, *Dark Paradise*, 46–86.

10. Courtwright, *Dark Paradise*, 46–86.

11. Courtwright, *Dark Paradise*, 46–86, 111.

12. Edward M. Brecher and the Editors of Consumer Reports, *Licit and Illicit Drugs* (Mount Vernon, NY: Consumers Union, 1972), 49–50.

13. Trebach, *The Heroin Solution*, 146–56.

14. Trebach, *The Heroin Solution*, 146–56.

15. Vincent Dole and Marie Nyswander, "Methadone Maintenance and Its Implication for Theories of Narcotic Addiction," *Research Publications of the Association for Research in Nervous and Mental Disease* 46 (1968): 359–66.

16. Vincent Dole and Marie Nyswander, *Methadone Maintenance: A Theoretical Perspective*, Theories on Drug Abuse NIDA Research Monograph 30 (Washington, DC: U.S. Government Printing Office, 1980), 256–61.

17. Snyder, *Brainstorming*, 9–19.

18. Trebach, *The Heroin Solution*, 233.

19. Lisa Sacco, *Drug Enforcement in the United States: History, Policy, and Trends in Illicit Drugs and Crime Policy*, October 2, 2014, Congressional Research Service 7-5700, https://fas.org/sgp/crs/misc/R43749.pdf.

20. Snyder, *Brainstorming*, 44–63.

21. Snyder, *Brainstorming*, 44–63, 120–57.

22. Carolyn Asbury, *Orphan Drugs: Medical Versus Market Value* (Lexington, MA: Lexington Book, D.C. Heath and Co., 1985), 61–64.

23. J. M. Perry, "Jack Anderson Empire Grows—and So Does Criticism It Receives. Column Is Called Reckless, Trivial, and He Decries Little Heed Paid to Him," *Wall Street Journal*, April 25, 1979.

24. Jack Anderson, "Drug Addicts: Unwilling Guinea Pigs," *Washington Post*, July 1, 1978.

25. Ted Gup and Jonathan Neumann, "Federal Contracts: A Litany of Frivolity, Waste," *Washington Post*, June 23, 1980.

26. Jerome Jaffe, "Can LAAM, Like Lazarus, Come Back from the Dead?" *Addiction* 102 (2007): 1342–43.

27. M. Douglas Anglin, Bradley T. Conner, Jeffery Annon, and Douglas Longshore, "Levo-Alpha-Acetylmethadol (LAAM) Versus Methadone Maintenance: 1-Year Treatment Retention, Outcomes and Status," *Addiction* 102 (2007): 1432–42.

28. H. Wieneke, H. Conrads, J. Wolstein, et al., "Levo-Alpha-Acetylmethadol (LAAM) Induced QTC-Prolongation—Results from a Controlled Clinical Trial," *European Journal of Medical Research* 14 (2009): 7–12.

29. Howard Sanders, "Drugs for Treating Narcotics Addicts," *Chemical & Engineering News*, March 28, 1977, 30–48.

30. Institute of Medicine, *Institute of Medicine (US) Committee on Federal Regulation of Methadone Treatment*, ed. R. A. Rettig and A. Yarmolinsky (Washington, DC: National Academies Press, 1995).

31. Cathie E. Alderks, "Trends in the Use of Methadone, Buprenorphine, and Extended-Release Naltrexone at Substance Abuse Treatment Facilities; 2003–2015 (UPDATE)," *CBHSQ Report*, 2017.

32. Anna Lembke and Jonathan Chen, "Use of Opioid Agonist Therapy for Medicare Patients in 2013," *Journal of the American Medical Association—Psychiatry* 73 (2016): 990–92.

33. Johann Hari, *Chasing the Scream: The First and Last Days of the War on Drugs* (New York: Bloomsbury, 2015), 231–55.

34. Hari, *Chasing the Scream*, 231–55.
35. Hari, *Chasing the Scream*, 231–55.
36. "In 1991, A Drug That Killed 17," *New York Times*, August 31, 1994.
37. Rudd et al., "Increases in Drug and Opioid Overdose Deaths."

23. The Toxicology of Climate Change

1. Mark Utell, "Effects of Inhaled Acid Aerosols on Lung Mechanics: An Analysis of Human Exposure Studies," *Environmental Health Perspectives* 63 (1985): 39–44.
2. Jane Koenig, David Covert, Quentin Hanley, Gerald Van Belle, and William Pierson, "Prior Exposure to Ozone Potentiates Subsequent Response to Sulfur Dioxide in Adolescent Asthmatic Subjects," *American Review of Respiratory Diseases* 141 (1990): 377–80.
3. Christine Corton, *London Fog: The Biography* (Cambridge, MA: Harvard University Press, 2015).
4. Robert Waller and Patrick Lawther, "Some Observations on London Fog," *British Medical Journal* 2 (1955): 1356–58.
5. W. P. D. Logan, "Mortality from Fog in London, January, 1956," *British Medical Journal* 1 (1956): 722–25.
6. Agency for Toxic Substances and Disease Registry, *Toxicological Profile for Mercury* (Atlanta, GA: Centers for Disease Control, 1999).
7. Alessandra Antunes dos Santos, Mariana Appel Hort, Megan Culbreth, et al., "Methylmercury and Brain Development: A Review of Recent Literature," *Journal of Trace Elements in Medicine and Biology* 38 (2016): 99–107.
8. Andrew Meiklejohn, "History of Lung Diseases of Coal Miners in Great Britain: I. 1800–1875," *British Journal of Industrial Medicine* 8 (1951): 127–37.
9. Mohsen Naghavi, Haidong Wang, Rafael Lozano, et al., "Global, Regional, and National Age-Sex Specific All-Cause and Cause-Specific Mortality for 240 Causes of Death, 1990–2013: A Systematic Analysis for the Global Burden of Disease Study," *Lancet* 385 (2013): 117–71.
10. Bernard Goldstein and Donald Reed, "Global Atmospheric Change and Research Needs in Environmental Health Sciences," *Environmental Health Perspectives* 96 (1991): 193–96.
11. Devra Davis, *When Smoke Ran Like Water* (New York: Basic Books, 2002), 260–69.
12. Devra Davis, "Short-Term Improvements in Public Health from Global-Climate Policies on Fossil-Fuel Combustion: An Interim Report. Working Group on Public Health and Fossil-Fuel Combustion," *Lancet* 350 (1997): 1341–49.
13. Ben Machol and Sarah Rizk, "Economic Value of U.S. Fossil Fuel Electricity Health Impacts," *Environment International* 52 (2013): 75–80.
14. Koenig et al., "Prior Exposure to Ozone Potentiates Subsequent Response to Sulfur Dioxide in Adolescent Asthmatic Subjects."
15. Nestor Molfino, Stanley Wright, Ido Katz, et al., "Effect of Low Concentrations of Ozone on Inhaled Allergen Responses in Asthmatic Subjects," *Lancet* 338 (1991): 199–203.
16. Sara Rasmussen, Elizabeth Ogburn, Meredith McCormack, et al., "Asthma Exacerbations and Unconventional Natural Gas Development in the Marcellus Shale," *Journal of the American Medical Association Internal Medicine* 176 (2016): 1334–43.

17. Office of Air Quality Planning and Standards, *Nitrogen Oxides (NOx), Why and How They Are Controlled*, EPA-456/F-99-006R (Research Triangle Park, NC: U.S. Environmental Protection Agency, 1999).

18. Jeff Deyette, Steven Clemmer, Rachel Cleetus, Sandra Sattler, Alison Bailie, and Megan Rising, *The Natural Gas Gamble: A Risky Bet on America's Clean Energy Future* (Cambridge, MA: Union of Concerned Scientists, 2015).

19. Ian Urbina, "Drilling Down: Regulation Lax as Gas Wells' Tainted Water Hits Rivers," *New York Times*, February 26, 2011.

20. Daniel Raimi, *The Fracking Debate: The Risks, Benefits, and Uncertainties of the Shale Revolution* (New York: Columbia University Press, 2018).

21. "OSHA-NIOSH Hazard Alert: Worker Exposure to Silica During Hydraulic Fracturing," https://www.osha.gov/dts/hazardalerts/hydraulic_frac_hazard_alert.html.

22. Committee on Energy and Commerce Minority Staff, *Chemicals Used in Hydraulic Fracturing* (Washington, DC: United States House of Representatives, 2011).

23. EPA, NAAQS table, https://19january2017snapshot.epa.gov/criteria-air-pollutants/naaqs-table_.html.

24. International Agency for Research on Cancer, "Diesel and Gasoline Engine Exhausts," in *IARC Monographs on the Evaluation of Carcinogenic Risks to Humans, Diesel and Gasoline Engine Exhausts and Some Nitroarenes* (Lyon: IARC, 2014).

25. M. Medina-Ramón, A. Zanobetti, J. Schwartz, "The Effect of Ozone and PM10 on Hospital Admissions for Pneumonia and Chronic Obstructive Pulmonary Disease: A National Multicity Study," *American Journal of Epidemiology* 163 (2006): 579–88.

26. W. Lawrence Beeson, David Abbey, and Synneve Knutsen, "Long-Term Concentrations of Ambient Air Pollutants and Incident Lung Cancer in California Adults: Results from the AHSMOG Study on Smog," *Environmental Health Perspectives* 106 (1998): 813–22.

27. Qian Di, Lingzhen Dai, Yun Wang, et al., "Association of Short-Term Exposure to Air Pollution with Mortality in Older Adults," *Journal of the American Medical Association* 318 (2017): 2446–56.

28. U.S. Environmental Protection Agency, *NATICH Data Base Report on State, Local, and EPA Air Toxics Activities*, EPA 450/3-90-012 (Washington, DC: U.S. Environmental Protection Agency, July 1990).

29. Adam Liptak and Coral Davenport, "Supreme Court Blocks Obama's Limits on Power Plants," *New York Times*, June 29, 2015.

30. Carolyn Brown and Rubin Thomerson, "United States Supreme Court Reverses Utility MACT Rule," *National Law Review* (September 20, 2016).

31. Lisa Friedman, "E.P.A. Proposal Puts Costs Ahead of Health Gains," *New York Times*, December 29, 2018.

32. John Parsons, Jacopo Buongiorno, Michael Corradini, and David Petti, "A Fresh Look at Nuclear Energy," *Science* 363 (2018): 105.

33. Elisabeth Cardis, Daniel Krewski, Mathieu Boniol, et al., "Estimates of the Cancer Burden in Europe from Radioactive Fallout from the Chernobyl Accident," *International Journal of Cancer* 119 (2006): 1224–35.

34. Cardis, Daniel Krewski, Mathieu Boniol, et al., "Estimates of the Cancer Burden in Europe."

35. Svetlana Alexievich, *Voices from Chernobyl: The Oral History of a Nuclear Disaster*, trans. Keith Gessen (New York: Picador, 2015), 129–34.

36. Benjamin Jones, "What Are the Health Costs of Uranium Mining? A Case Study of Miners in Grants, New Mexico," *International Journal of Occupational and Environmental Health* 20 (2014): 289–300.

37. Robert Service, "Advances in Flow Batteries Promise Cheap Backup Power," *Science* 362 (2018): 508–9.

38. Keith Bradsher, "In China, Illegal Rare Earth Mines Face Crackdown," *New York Times*, December 29, 2010.

39. Dustin Mulvaney, "Solar Energy Isn't Always as Green as You Think. Do Cheaper Photovoltaics Come with a Higher Environmental Price Tag?" *IEEE Spectrum*, August 26, 2014.

40. World Health Organization, "Ten Threats to Global Health," https://www.who.int/emergencies/ten-threats-to-global-health-in-2019.

41. Damian Carrrington and Matthew Taylor, "Air Pollution Is the 'New Tobacco,' Warns WHO Head," *Guardian*, October 27, 2018.

24. Animal Models for Human Disease

1. Sudhir Kumar and S. Blair Hedges, "A Molecular Timescale for Vertebrate Evolution," *Nature* 392 (1998): 917–20.

2. Robert Perlman, "Mouse Models of Human Disease: An Evolutionary Perspective," *Evolution Medicine and Public Health* 1 (2016): 170–76.

3. Annapoorni Rangarajan and Robert A. Weinberg, "Opinion: Comparative Biology of Mouse Versus Human Cells: Modeling Human Cancer in Mice," *Nature Reviews Cancer* 3 (2003): 952–59.

4. Gordon Hard and Kanwar Khan, "A Contemporary Overview of Chronic Progressive Nephropathy in the Laboratory Rat, and Its Significance for Human Risk Assessment," *Toxicologic Pathology* 32 (2004): 171–80.

5. Joseph Haseman, James Hailey, and Richard Morris, "Spontaneous Neoplasm Incidences in Fischer 344 Rats and B6C3F1 Mice in Two-Year Carcinogenicity Studies: A National Toxicology Program Update," *Toxicologic Pathology* 26 (1998): 428–41.

6. Hui Huang, Eitan Winter, Huajun Wang, et al., "Evolutionary Conservation and Selection of Human Disease Gene Orthologs in the Rat and Mouse Genomes," *Genome Biology* 5 (2004): 47.

7. Mick Bailey, Zoe Christoforidou, and Marie Lewis, "The Evolutionary Basis for Differences Between the Immune Systems of Man, Mouse, Pig, and Ruminants," *Veterinary Immunology and Immunopathology* 152 (2013): 13–19.

8. John Whysner, Carson Conaway, Lynn Verna, and Gary Williams, "Vinyl Chloride Mechanistic Data and Risk Assessment: DNA Reactivity and Cross-Species Quantitative Risk Extrapolation," *Pharmacology and Therapeutics* 71 (1996): 7–28.

9. William Hahn and Robert Weinberg, "Modelling the Molecular Circuitry of Cancer," *Nature Reviews Cancer* 2 (2002): 331–41.

10. Hahn and Weinberg, "Modelling the Molecular Circuitry of Cancer."

11. Harry Olson, Braham Betton, Denise Robinson, et al., "A Concordance of the Toxicity of Pharmaceuticals in Humans and in Animals," *Regulatory Toxicology Pharmacology* 32 (2000): 56–67.

12. European Commission "Of Mice and Men—Are Mice Relevant Models for Human Disease? Outcomes of the European Commission Workshop 'Are Mice Relevant Models for Human Disease?' Held in London, UK, on 21 May 2010," http://ec.europa.eu/research/health/pdf/summary-report-25082010_en.pdf.

13. Amelia Kellar, Cay Egan, and Don Morris, "Preclinical Murine Models for Lung Cancer: Clinical Trial Applications," *BioMed Research International* (2015): doi:10.1155/2015/621324.

14. John Minna, Jonathan Kurie, and Tyler Jacks, "A Big Step in the Study of Small Cell Lung Cancer," *Cancer Cell* 4 (2003): 163–66.

15. Robert Kerbel, "What Is the Optimal Rodent Model for Anti-Tumor Drug Testing?" *Cancer Metastasis Review* 17 (1998–1999): 301–4; Kenneth Paigen, "One Hundred Years of Mouse Genetics: An Intellectual History: II. The Molecular Revolution (1981–2002)," *Genetics* 163 (2003): 1227–35.

16. Judah Folkman, Ezio Merler, Charles Abernathy, and Gretchen Williams, "Isolation of a Tumor Factor Responsible for Angiogenesis," *Journal of Experimental Medicine* 133 (1971): 275–88.

17. Robert Langer, Howard Conn, Joseph Vacanti, Christian Haudenschild, and Judah Folkman, "Control of Tumor Growth in Animals by Infusion of an Angiogenesis Inhibitor," *Proceedings of the National Academy of Sciences, USA* 77 (1980): 4331–35.

18. Gina Kolata, "Hope in the Lab: A Special Report. A Cautious Awe Greets Drugs That Eradicate Tumors in Mice," *New York Times*, May 3, 1998.

19. Joe Nocera, "Why Doesn't No Mean No?" *New York Times*, November 21, 2011.

20. Jean Marx, "Angiogenesis: A Boost for Tumor Starvation," *Science* 301 (2003): 452–54.

21. Judah Folkman, "Antiangiogenesis in Cancer Therapy—Endostatin and Its Mechanisms of Action," *Experimental Cell Research* 312 (2006): 594–607.

22. Folkman, "Antiangiogenesis in Cancer Therapy."

23. Jocelyn Kaiser, "The Cancer Test," *Science* 348 (2015): 1411–14.

24. Martin Enserink, "Sloppy Reporting on Animal Studies Proves Hard to Change," *Science* 357 (2017): 1337–38.

25. Kelly Servick, "Of Mice and Microbes," *Science* 353 (2016): 741–43.

26. Perlman, "Mouse Models of Human Disease."

25. Are Animal Cancer Bioassays Reliable?

1. David Rall, "The Role of Laboratory Animal Studies in Estimating Carcinogenic Risks for Man," IARC Scientific Publication 25 (Lyon: IARC, 1979), 179–89.

2. According to the IARC, other carcinogens are radionucleotides or infectious agents, and some workplace environments are considered carcinogenic.

3. "Agents Classified by the IARC Monographs, Volumes 1–124," http://monographs.iarc.fr/ENG/Classification/index.php. There are other kinds of agents, such as radiation, biologicals, and working environments, but only forty to fifty distinct chemicals.

4. International Agency for Research on Cancer, "Arsenic in Drinking Water," in *Some Drinking Water Disinfectants and Contaminants, Including Arsenic*, IARC Monographs on the Evaluation of Carcinogenic Risk of Chemicals to Humans 84 (Lyon: IARC, 2004), 41–267.

5. International Agency for Research on Cancer, "Arsenic and Arsenic Compounds," in *Arsenic, Metals, Fibres, and Dusts*, IARC Monographs on the Evaluation of Carcinogenic Risk of Chemicals to Humans 100C (Lyon: IARC, 2012), 41–93.

6. Joseph Haseman and Ann-Marie Lockhart, "Correlations Between Chemically Related Site-Specific Carcinogenic Effects in Long-Term Studies in Rats and Mice," *Environmental Health Perspectives* 101 (1993): 50–54.

7. Sudhir Kumar and S. Blair Hedges, "A Molecular Timescale for Vertebrate Evolution," *Nature* 392 (1998): 917–20.

8. Manik Chandra and Charles Frith, "Spontaneous Neoplasms in Aged CD-1 Mice," *Toxicology Letters* 61 (1992): 67–74. Joseph Haseman, James Hailey, and Richard Morris, "Spontaneous Neoplasm Incidences in Fischer 344 Rats and B6C3F1 Mice in Two-Year Carcinogenicity Studies: A National Toxicology Program Update," *Toxicologic Pathology* 26 (1998): 428–41.

9. Bruce Ames and Lois Gold, "Chemical Carcinogenesis: Too Many Rodent Carcinogens," *Proceedings of the National Academy of Sciences, USA* 87 (1990): 7772–76; Lois Gold, Neela Manley, Thomas Slone, Georganne Garfinkel, Lars Rohrbach, and Bruce Ames, "The Fifth Plot of the Carcinogenic Potency Database: Results of Animal Bioassays Published in the General Literature Through 1988 and by the National Toxicology Program Through 1989," *Environmental Health Perspectives* 100 (1993): 65–168.

10. E. Gottmann, S. Kramer, B. Pfahringer, and C. Helma, "Data Quality in Predictive Toxicology: Reproducibility of Rodent Carcinogenicity Experiments," *Environmental Health Perspectives* 109 (2001): 509–14.

11. International Agency for Research on Cancer, "Reserpine," in *Some Pharmaceutical Drugs*, IARC Monographs on the Evaluation of Carcinogenic Risks to Humans 24 (Lyon: IARC, 1980), 211–41.

12. International Agency for Research on Cancer, *Peroxisome Proliferation and Its Role in Carcinogenesis*, IARC Technical Report 24 (Lyon: IARC, 1995).

13. International Agency for Research on Cancer, "Clofibrate," in *Some Pharmaceutical Drugs*, IARC Monographs on the Evaluation of Carcinogenic Risks to Humans 66 (Lyon: IARC, 1996), 391–426.

14. J. Christopher, Michael Corton, B. Cunningham, et al. "Mode of Action Framework Analysis for Receptor-Mediated Toxicity: The Peroxisome Proliferator-Activated Receptor Alpha (Pparα) as a Case Study," *Critical Reviews of Toxicology* 44 (2014): 1–49.

15. Gordon Hard and John Whysner, "Risk Assessment of D-Limonene: An Example of Male Rat-Specific Renal Tumorigens," *Critical Reviews of Toxicology* 24 (1994): 231–54.

16. O. G. Fitzhugh, A. A. Nelson, and J. P. Frawley, "A Comparison of the Chronic Toxicities of Synthetic Sweetening Agents," *Journal of the American Pharmaceutical Association* 40 (1951): 583–86.

17. John Whysner and Gary Williams, "Saccharin Mechanistic Data and Risk Assessment: Urine Composition, Enhanced Cell Proliferation, and Tumor Promotion," *Pharmacology & Therapeutics* 71 (1996): 225–52.

18. Samuel Cohen, Martin Cano, Robert Earl, Stephen Carson, and Emily Garland, "A Proposed Role for Silicates and Protein in the Proliferative Effects of Saccharin on the Male Rat Urothelium," *Carcinogenesis* 12 (1991): 1551–55.

19. International Agency for Research on Cancer, "Saccharin and Its Salts," in *Some Chemicals That Cause Tumours of the Kidney or Urinary Bladder in Rodents and Some*

Other Substances, IARC Monographs on the Evaluation of Carcinogenic Risks to Humans 73 (Lyon: IARC, 1999), 517–624.

20. American Cancer Society, Cancer Statistics Center, https://cancerstatisticscenter .cancer.org/?_ga=1.174306402.1115391928.1446921311#/.

21. American Cancer Socety, "Liver Cancer Risk Factors," http://www.cancer.org /cancer/livercancer/detailedguide/liver-cancer-risk-factors.

22. Girard Hottendorf and Irwin Pachter, "Review and Evaluation of the NCI/NTP Carcinogenesis Bioassays," *Toxicologic Pathology* 13 (1985): 141–46.

23. Jorgen Olsen, Gabi Schulgen, John Boice Jr., et al., "Antiepileptic Treatment and Risk for Hepatobiliary Cancer and Malignant Lymphoma," *Cancer Research* 55 (1995): 294–97.

24. Mariana Pereira, "Route of Administration Determines Whether Chloroform Enhances or Inhibits Cell Proliferation in the Liver of B6C3F1 Mice," *Fundamental and Applied Toxicology* 23 (1994): 87–92.

25. EPA, "Chloroform," https://cfpub.epa.gov/ncea/iris2/chemicalLanding.cfm?substance _nmbr=25.

26. M. E. (Bette) Meek, John Bucher, Samuel Cohen, et al., "A Framework for Human Relevance Analysis of Information on Carcinogenic Modes of Action," *Critical Reviews in Toxicology* 33 (2003): 591–653.

27. Robert Maronpot, Abraham Nyskab, Jennifer Foremanc, and Yuval Ramotd, "The Legacy of the F344 Rat as a Cancer Bioassay Model (A Retrospective Summary of Three Common F344 Rat Neoplasms)," *Critical Reviews in Toxicology* 46 (2016): 641–75.

26. Hormone Mimics and Disrupters

1. Elizabeth Watkins, *The Estrogen Elixir: A History of Hormone Replacement Therapy in America* (Baltimore, MD: Johns Hopkins University Press, 2007), 12–13; H. M. Bolt, "Metabolism of Estrogens—Natural and Synthetic," *Pharmacology and Therapeutics* (1979): 155–81.

2. Evon Simpson and Richard Santen, "Celebrating 75 Years of Oestradiol," *Journal of Molecular Endocrinology* 55 (2015): T1–20.

3. Michael Shimkin, *Contrary to Nature: Being an Illustrated Commentary on Some Persons and Events of Historical Importance in the Development of Knowledge Concerning . . . Cancer* (Washington, DC: Department of Health Education and Welfare, 1977), DHEW publication no. (NIH) 76-720, chap. VIII-F-1.

4. International Agency for Research on Cancer, "Diethylstilboestrol and Diethylstilboestrol Dipropionate," in *Sex Hormones*, IARC Monographs on the Evaluation of Carcinogenic Risks to Humans 21 (Lyon: IARC, 1979), 173–231.

5. IARC, "Diethylstilboestrol and Diethylstilboestrol Dipropionate."

6. International Agency for Research on Cancer, "Post-Menopausal Oestrogen Therapy," in *Hormonal Contraception and Post-Menopausal Hormonal Therapy*, IARC Monographs on the Evaluation of Carcinogenic Risks to Humans 72 (Lyon: IARC, 1999), 399–400.

7. Robert Hoover, Laman Gray Sr., Phillip Cole, and Brian MacMahon, "Menopausal Estrogens and Breast Cancer," *New England Journal of Medicine* 295 (1976): 401–5.

8. IARC, "Post-Menopausal Oestrogen Therapy."

9. Elwood Jensen, "From Chemical Warfare to Breast Cancer Management," *Nature Medicine* 10 (2004): 1018–21.

10. Ana Soto, James Murai, Pentti Siiteri, and Carlos Sonnenschein, "Control of Cell Proliferation: Evidence for Negative Control on Estrogen-Sensitive T47D Human Breast Cancer Cells," *Cancer Research* 46 (1986): 2271–75.

11. Theo Colborn, Dianne Dumanoski, and John Peterson Myers, *Our Stolen Future* (New York: Plume, 1997), 122–41.

12. Ana Soto, Honorato Justicia, Jonathan Wray, and Carlos Sonnenschein, "P-Nonylphenol: An Estrogenic Xenobiotic Released from 'Modified' Polystyrene," *Environmental Health Perspectives* 92 (1991): 167–73.

13. Ana Soto, T-M Lin, H. Justicia, R. M. Silvia, and Carlos Sonnenschein, "An 'in Culture' Bioassay to Assess the Estrogenicity of Xenobiotics (E-SCREEN)," in *Chemically Induced Alterations in Sexual and Functional Development: The Wildlife/Human Connection*, ed. Theo Colborn and Coralie Clement (Princeton, NJ: Princeton Scientific Publishing Co., 1992), 295–309.

14. A.M. Soto, C. Sonnenschein, K. L. Chung, M. F. Fernandez, N. Olea, and F. O. Serrano, "The E-SCREEN Assay as a Tool to Identify Estrogens: An Update on Estrogenic Environmental Pollutants," *Environmental Health Perspectives* 103 (1995): 113–22.

15. Colborn, Dumanoski, and Myers, *Our Stolen Future*, 122–41.

16. Robert Golden, Kenneth Noller, Linda Titus-Ernstoff, et al., "Environmental Endocrine Modulators and Human Health: An Assessment of the Biological Evidence," *Critical Reviews of Toxicology* 28 (1998): 109–227.

17. Robert Nilsson, "Endocrine Modulators in the Food Chain and Environment," *Toxicologic Pathology* 28 (2000): 420–31.

18. Eugenia Calle, Howard Frumkin, S. Jane Henley, David Savitz, and Michael Thun, "Organochlorines and Breast Cancer Risk," *CA: A Cancer Journal for Clinicians* 52 (2002): 301–9.

19. International Agency for Research on Cancer, "Atrazine," in *Some Chemicals That Cause Tumours of the Kidney or Urinary Bladder in Rodents and Some Other Substances*, IARC Monographs on the Evaluation of Carcinogenic Risks to Humans 73 (Lyon: IARC, 1999), 59–113.

20. Norman Adams, "Detection of the Effects of Phytoestrogens on Sheep and Cattle," *Journal of Animal Sciences* 73 (1995): 1509–15; K. F. M. Reed, "Fertility of Herbivores Consuming Phytoestrogen-Containing Medicago and Trifolium Species," *Agriculture* 6 (2016): 35.

21. Christopher Borgert, John Matthews, and Stephan Baker, "Human-Relevant Potency Threshold (HRPT) for ERα Agonism," *Archives of Toxicology* 92 (2018): 1685–1702.

22. National Research Council (U.S.) Committee on Hormonally Active Agents in the Environment, *Hormonally Active Agents in the Environment* (Washington, DC: National Academies Press, 1999).

23. Marcello Cocuzza and Sandro Esteves, "Shedding Light on the Controversy Surrounding the Temporal Decline in Human Sperm Counts: A Systematic Review," *Scientific World Journal* 2 (2014): Article ID 365691; Allen Pacey, "Are Sperm Counts Declining? Or Did We Just Change Our Spectacles?" *Asian Journal of Andrology* 15 (2013):187–90.

24. Hagai Levine, Niels Jørgensen, Anderson Martino-Andrade, et al., "Temporal Trends in Sperm Count: A Systematic Review and Meta-Regression Analysis," *Human Reproduction Update* 23 (2017): 646–59.

25. Anne Vested, Cecilia Ramlau-Hansen, Sjurdur Olsen, et al. "In Utero Exposure to Persistent Organochlorine Pollutants and Reproductive Health in the Human Male," *Reproduction* 148 (2014): 635–46.

26. Golden et al., "Environmental Endocrine Modulators and Human Health."

27. Raphael Witorsch, "Endocrine Disruptors: Can Biological Effects and Environmental Risks Be Predicted?" *Regulatory Toxicology and Pharmacology* 36 (2002): 118–30.

28. Christopher Borgert, E. V. Sargent, G. Casella, et al., "The Human Relevant Potency Threshold: Reducing Uncertainty by Human Calibration of Cumulative Risk Assessments," *Regulatory Toxicology and Pharmacology* 62 (2012): 313–28.

29. International Agency for Research on Cancer, "General Remarks," in *Some Thyrotropic Agents*, IARC Monographs on the Evaluation of Carcinogenic Risks to Humans 79 (Lyon: IARC, 2001), 33–46.

30. International Agency for Research on Cancer, "Sulfamethazine and its Sodium Salt," in *Some Thyrotropic Agents*, IARC Monographs on the Evaluation of Carcinogenic Risks to Humans 79 (Lyon: IARC, 2001), 341–60.

27. Building Better Tools for Testing

1. Unnur Thorgeirsson, Dan Dalgard, Jeanette Reeves, and Richard Adamson, "Tumor Incidence in a Chemical Carcinogenesis Study of Nonhuman Primates," *Regulatory Toxicology and Pharmacology* 19 (1994): 130–51.

2. David Eastmond, Suryanarayana Vulimiri, John French, and Babasaheb Sonawane, "The Use of Genetically Modified Mice in Cancer Risk Assessment: Challenges and Limitations," *Critical Reviews of Toxicology* 43 (2013): 611–31.

3. Eastmond et al., "The Use of Genetically Modified Mice in Cancer Risk Assessment."

4. Joyce McCann, Edmond Choi, Edith Yamasaki, and Bruce Ames, "Detection of Carcinogens as Mutagens in the Salmonella/Microsome Test: Assay of 300 Chemicals," *Proceedings of the National Academy of Science, USA* 72 (1975): 5135–39.

5. I. F. Purchase, E. Longstaff, John Ashby, et al., "An Evaluation of 6 Short-Term Tests for Detecting Organic Chemical Carcinogens," *British Journal of Cancer* 37 (1978): 873–903.

6. David Kirkland, Errol Zeiger, Federica Madiac, and Raffaella Corvic, "Can In Vitro Mammalian Cell Genotoxicity Test Results Be Used to Complement Positive Results in the Ames Test and Help Predict Carcinogenic or In Vivo Genotoxic Activity? II. Construction and Analysis of a Consolidated Database," *Mutation Research/Genetic Toxicology and Environmental Mutagenesis* 775–776 (2014): 69–80.

7. Errol Zeiger, "Identification of Rodent Carcinogens and Noncarcinogens Using Genetic Toxicity Tests: Premises, Promises, and Performance," *Regulatory Toxicology and Pharmacology* 28 (1998): 85–95.

8. David Kirkland, Marilyn Aardem, Leigh Henderson, and Lutz Muller, "Evaluation of the Ability of a Battery of Three In Vitro Genotoxicity Tests to Discriminate Rodent Carcinogens and Non-Carcinogens I. Sensitivity, Specificity, and Relative Predictivity," *Mutation Research* 584 (2005): 1–256.

9. John Weisburger and Gary Williams, "The Distinct Health Risk Analyses Required for Genotoxic Carcinogens and Promoting Agents," *Environmental Health Perspectives* 50 (1983): 233–45.

10. Committee on Toxicity Testing and Assessment of Environmental Agents, National Research Council, *Toxicity Testing in the 21st Century: A Vision and a Strategy* (Washington, DC: National Academies Press, 2007).

11. Jill Niemark, "Line of Attack: Christopher Korch Is Adding Up the Cost of Contaminated Cell Lines," *Science* 347 (2015): 938–40.

12. Liwen Vaughan, Wolfgang Glanzel, Christopher Korch, and Amanda Capes-Davis, "Widespread Use of Misidentified Cell Line KB (HeLa): Incorrect Attribution and Its Impact Revealed Through Mining the Scientific Literature," *Cancer Research* 77 (2017): 2784–88.

13. Niemark, "Line of Attack."

14. Katie Friedman, Eric Watt, Michael Hornung, et al., "Tiered High-Throughput Screening Approach to Identify Thyroperoxidase Inhibitors Within the ToxCast Phase I and II Chemical Libraries," *Toxicological Sciences* 151 (2016): 160–80.

15. Jeremy Leonard, Yu-Mei Tan, Mary Gilbert, Kristin Isaacs, and Hisham El-Masri, "Estimating Margin of Exposure to Thyroid Peroxidase Inhibitors Using High-Throughput In Vitro Data, High-Throughput Exposure Modeling, and Physiologically Based Pharmacokinetic/Pharmacodynamic Modeling," *Toxicological Sciences* 151 (2016): 57–70.

28. An Ounce of Prevention Is Worth a Pound of Cure

1. Rachel Carson, *Silent Spring* (Boston: Houghton Mifflin Company, 1962), 240–43.

2. Carson, *Silent Spring*, 58–59.

3. Victor Ekpu and Abraham Brown, "The Economic Impact of Smoking and of Reducing Smoking Prevalence: Review of Evidence," *Tobacco Use Insights* 8 (2015): 1–35.

4. Tobacco Free Kids, https://www.tobaccofreekids.org.

5. Elise Gould, "Childhood Lead Poisoning: Conservative Estimates of the Social and Economic Benefits of Lead Hazard Control," *Environmental Health Perspectives* 117 (2009): 1162–67.

6. National Institute of Drug Abuse, "Trends and Statistics," https://www.drugabuse.gov/related-topics/trends-statistics.

7. Marcella Boynton, Robert Agans, J. Michael Bowling, et al., "Understanding How Perceptions of Tobacco Constituents and the FDA Relate to Effective and Credible Tobacco Risk Messaging: A National Phone Survey of U.S. Adults, 2014-2015," *BMC Public Health* 16 (2016): 516.

8. U.S. Surgeon General, National Center for Chronic Disease Prevention and Health Promotion (US) Office on Smoking and Health, *The Health Consequences of Smoking—50 Years of Progress: A Report of the Surgeon General* (Atlanta, GA: Centers for Disease Control and Prevention, 2014).

9. Surgeon General, *The Health Consequences of Smoking*.

10. Surgeon General, *The Health Consequences of Smoking*.

11. Boynton et al., "Understanding How Perceptions of Tobacco Constituents and the FDA Relate to Effective and Credible Tobacco Risk Messaging."

12. Martin Makary and Michael Daniel, "Medical Error--The Third Leading Cause of Death in the US," *British Medical Journal* 353 (2016): i2139.

13. Sabrina Tavernise, "Smokers Urged to Switch to E-Cigarettes by British Medical Group," *New York Times*, April 27, 2016; Sabrina Tavernise, "Safer to Puff, E-Cigarettes Can't Shake Their Reputation as a Menace," *New York Times*, November 1, 2016.
14. Internal Agency for Research on Cancer, "Tobacco Smoking," in *Personal Habits and Indoor Combustions*, Monographs on the Evaluation of Carcinogenic Risks to Humans 100E (Lyon: IARC, 2012), 43–211.
15. Lorenzo Tomatis, "Identification of Carcinogenic Agents and Primary Prevention of Cancer," *Annals of the New York Academy of Science* 1076 (2006): 1–14.

Index